D0919504

Production Enhancement *with* Acid Stimulation

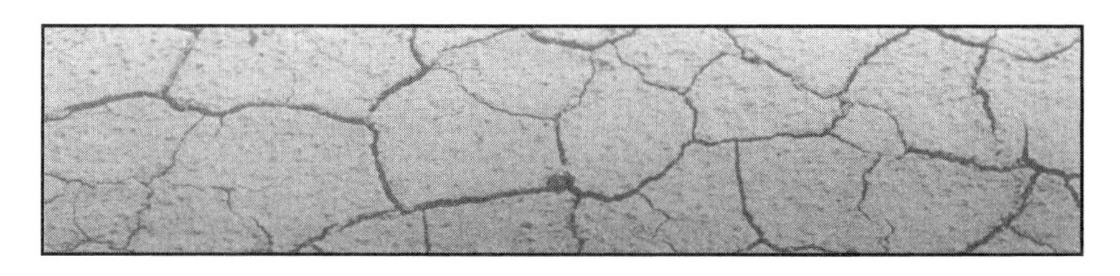

Production Enhancement *with* Acid Stimulation

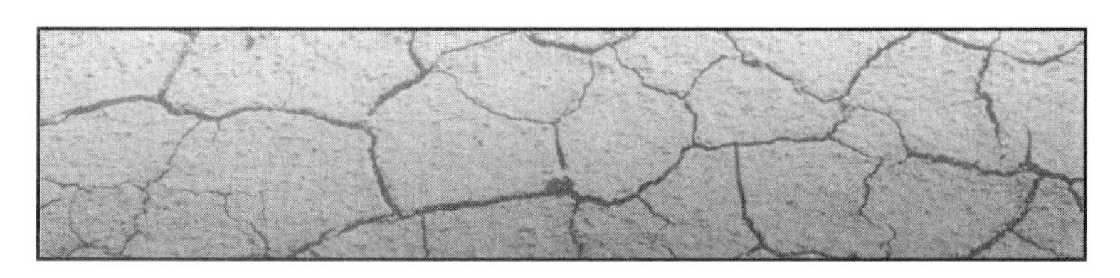

By Leonard Kalfayan

Copyright © 2000 by
PennWell Corporation
1421 South Sheridan/P.O. Box 1260
Tulsa, Oklahoma 74112/74101

ISBN 0-87814-778-0

Printed in the United States of America

All rights reserved. No part of this book may be reproduced, stored in a retrieval system, or transcribed in any form or by any means, electronic or mechanical, including photocopying and recording, without the prior written permission of the publisher.

DEDICATION

To my father, Dr. Sarkis Kalfayan, who was a true scientist, and to my mother, Irene Kalfayan, who always encourages me.

CONTENTS

PART IV. SPECIALIZED REMEDIAL TREATMENTS

PART V. QUALITY CONTROL PRACTICES

APPENDICES

LIST OF FIGURES

LIST OF TABLES

Introduction

The 100-year history of acidizing is filled with discovery and ingenuity, spectacular success and disappointing failure. Its potential for prolific well stimulation has always been tempered by its unpredictability and by the continuing frustration it seems to cause. It cannot and will not be defined by an exact set of rules and expectations.

However, in terms of potential return on investment and generation of immediate productivity enhancement and cash flow, at a reasonable price, acidizing has no equal. Unfortunately, this tremendous attribute is largely unappreciated. Acidizing is often perceived as a treatment of last resort, or as a necessary evil when it comes to remedial well service. It is not often perceived as the high-return, often low-cost well stimulation process that it is.

This lack of appreciation for acidizing partly stems from the natural and predominant oilfield preference for hydraulic fracturing, which is, admittedly, more appealing to the masses. What is unfortunate, though, is that acidizing often is viewed incorrectly as a competing method, rather than as a viable optional addition to hydraulic fracturing in the stimulation "toolbox."

The common frustration with acidizing results mostly from the unending desire to impart predictability through forced systematization, in order to develop reproducible treatment designs. This approach can work with fracturing in certain areas. However, with acidizing, history has shown repeatedly that this approach is doomed to failure and to certain discouragement.

What history has shown is that acidizing is not an exact, predictable science. Acidizing is an inexact science, as well as an art that cannot be completely mastered. (Some would call it a black art, but it is not mysterious to that extent any more.) In any case, all of the arts are sciences, and much of science is art too, though not always recognized as such. For example, music is both an art and a complex mathematical science. Proficiency and enjoyment result as a feel for the art is developed, not from a full understanding of its complexities. There is always opportunity for experimentation, new concepts, and new approaches.

The same can be said of acidizing. Complete mastery is not possible, but it is also not necessary for success. And there can be excitement and satisfaction in the risk-taking and creative possibilities, and in the tremendous profit potential that acidizing stimulation affords.

There are many written sources on acidizing and related subjects. There are excellent texts covering topics related to stimulation, including acidizing. There have also been many articles and technical papers on acidizing in the industry literature. Most of these are sanctioned and published by the Society of Petroleum Engineers (SPE) in meeting proceedings and in society journals.

This information includes well and field stimulation case histories, specific aspects of acidizing fundamentals and research, and development of new and improved methods, products, and novel applications. Many such contributions to the industry are made each year, and some are meaningful. In addition, there are in-house company guidelines, "best practices," and the like. There is also in-house information available in service companies, oil companies, and generally throughout the industry.

This book is neither an attempt to supply the industry with a textbook on acidizing, nor is it a comprehensive compilation of technical contributions to the industry in subjects related to acidizing. It is also not a simplified acidizing treatment design "cookbook."

History has shown repeatedly that acidizing cannot be and should not be subject to cookbook procedures. Only certain steps in the design decision process may be treated as such, as we shall see. With acidizing, there are many more exceptions to the rules than there are rules. In fact, true success in acidizing is associated with the better understanding of the exceptions.

This book represents an interest in promoting an appreciation for the art of acidizing. Its contents are intended to serve as a practical guide when considering the stimulation decision process, acid treatment design, and acid treatment benefits and limitations in both carbonate and sandstone formations.

It is really a commentary on what I consider to be of practical importance with respect to acid treatment design. Certain portions are inspired by the work of George King and Harry McLeod. McLeod, in particular, has contributed to the understanding and practice of acid stimulation, especially in sandstones. It is additionally influenced by others, including long-time colleague Matt Zielinski, Joel Boles, David Watkins, Roland Krueger, Rick Gdanski, and Syed Ali.

Beyond what is presented here, I hope that you find the interest and curiosity to explore the texts and the literature. These will allow you to learn more about this remarkable and extremely profitable, yet often misunderstood and unappreciated, area of the well stimulation industry.

I further hope that your continued research and investigation include those works presenting acidizing history, acidizing mathematics, physics, and chemistry, as well as the latest popular acidizing methods, applications, and products. But, I hope that you can consider and separate only those that are valid and useful. I hope this book helps you to do so, and helps you to develop your own feel for acidizing and its possibilities.

part one

preliminaries

A Brief History of Acidizing 1

The history of acidizing is remarkable, and the story from the 1890s through the 1960s is detailed beautifully in the first chapter of the Society of Petroleum Engineers (SPE) monograph, *Acidizing Fundamentals*, by Williams, Gidley, and Schechter.[1] Unfortunately, the monograph is difficult to come by. Therefore this contribution to the literature is included in summary here, with some additional commentary.

CARBONATE ACIDIZING

Acidizing predates just about all well stimulation techniques. Other techniques, such as hydraulic fracturing, were developed much more recently. Acidizing may, in fact, be the oldest stimulation technique still in modern use. The earliest acid treatments of oil wells are believed to have occurred as far back as 1895.[2] The Standard Oil Company used concentrated hydrochloric acid (HCl) to stimulate oil wells producing from carbonate formations in Lima, Ohio, at their Solar Refinery.

Herman Frasch, the Solar Refinery chief chemist at that time, is credited with the invention of the acidizing technique.[3] Frasch was issued the first patent on acidizing on March 17, 1896.[4] The brief Frasch patent was the first of many acidizing patents. In the patent, Frasch proposed commercial muriatic acid (30–40% by weight of HCl, a highly water-soluble gaseous acid).

A similar patent was granted to John Van Dyke, general manager of the Solar Refinery at that time, using sulfuric acid rather than hydrochloric acid. Van Dyke was Frasch's close associate and boss. Each assigned half an interest in each other's patents—perhaps indicating that neither knew which process might be more successful. Or perhaps it was a matter of management seizing an opportunity to take credit, at least partially, for a potentially significant invention.

As it turned out, the Frasch patent was the more successful. Hydrochloric acid would react in limestone to form the soluble products carbon dioxide (CO_2) and calcium chloride ($CaCl_2$). These products could be produced out of the formation once a treated well was returned to production. In contrast, sulfuric acid (H_2SO_4) produces insoluble calcium sulfate, which could plug the formation. Therefore Frasch is credited with the invention of acidizing.

The Frasch patent was not only the first acidizing patent, but the most instructive, as well. Many of the elements of modern-day acidizing are included in the Frasch patent. Frasch suggested the need to "put the acid under strong pressure" so that it might be "pressed into the rock and made to act upon the same at a distance from the original well-hole." He also mentioned that "long channels can be formed." He also recommended the use of an overflush, stating, "It is advantageous to displace it (acid) and cause it to penetrate further into the rock by forcing a neutral or cheap liquid, such as water, into the well."

The need for corrosion protection was also anticipated. Frasch proposed "to introduce an alkaline liquid (preferably milk of lime)" to neutralize any unspent acid returned to the wellbore following treatment. The advantage of neutralization, he said, "is to avoid the danger of corroding the subsequently used apparatus." Frasch proposed the use of pipe that was "enameled or lead lined" or to "otherwise make proof against corrosion." Finally, Frasch pointed to the need for a rubber packer to isolate the zone and force the acid into the formation to be treated.

The acidizing process was applied with great success in the Lima, Ohio wells. Many wells were acidized with remarkable results in the short term. However, its use soon declined, and acidizing was used very infrequently during the next 30 years or so. It is not clear why that was the case. Perhaps it was related to the lack of an effective method for limiting acid corrosion. However, throughout its history, acidizing has a repeating record of quickly and inexplicably losing popularity, seemingly independent of results at times.

SCALE REMOVAL

In 1928, the use of acidizing rose again, albeit briefly. The Gypsy Oil Company, a subsidiary of Gulf Oil Company, used hydrochloric acid to remove calcareous scale deposited in the pipe and equipment in wells in Oklahoma.[5] Treatment recommendations for these wells were provided by Dr. Blain Westcott of the Mellon Institute at the request of Gypsy Oil. Interestingly enough, his recommendations included an acid corrosion inhibitor, Rodine No. 2, which was an inhibitor used in acid pickling in steel mills.

A patent application on the use of corrosion inhibitor in well stimulation application was not filed. Apparently this was because it was thought to be an old art adapted from the steel industry, dating back as far as about 1845.[6] In any case, the treatments were successful. However, its use declined in the early 1930s with a drop in oil prices.[5]

With the eventual development of a suitable commercial corrosion inhibitor (arsenic) by the Dow Chemical Company in the early 1930s, use of the acidizing process finally escalated. In 1932, the Pure Oil Company and Dow Chemical collaborated to successfully stimulate several oil wells in Michigan limestone formations with hydrochloric acid treatments. Pure had oil property in Michigan and an active exploration program there. Dow had brine wells in the area. Therefore, Pure asked Dow to provide operational assistance.

Since Dow had no interest in oil well production, the company agreed to make its brine well files available to Pure. In discussions between Pure and Dow, it was decided to treat the limestone formations with hydrochloric acid. Pure suggested HCl, based on its reactivity with limestone. As it happened, Dow had some experience treating its brine wells, which produced from sandstone formations, with HCl.

It appears from the historical record that the principals were unaware of the earlier work by Frasch and Standard Oil. In any case, Pure decided to acidize one of its own wells—the Fox No. 6 well in Section 13, Chippewa Township, Isabella County, Michigan, on February 11, 1932. As written in *Acidizing Fundamentals* (which the authors drew from several sources[7–9]):

> *The well was treated with 500 gallons of hydrochloric acid. Acid was brought to the well site on a tank wagon equipped with a wooden tank 36 inches in diameter and 12 feet long. To this acid, 2 gallons of an arsenic acid inhibitor were added, at the suggestion of John Grebe, head of Dow's Physical Research Laboratory, to reduce corrosion of the tubing. The acid was transferred from the tank truck to the wellbore by siphoning with a garden hose. About half of the 500 gallons of acid was siphoned into the tubing. This was followed by 6 bbl of oil pumped into the tubing (with a hand-operated pump) after the acid. The well was shut in over night and swabbed the next morning. A large quantity of emulsion was removed. The remaining acid was siphoned into the tubing and displaced by oil flush. The well, which was dead before treatment, produced as much as 16 barrels of oil per day afterwards. Other wells were subsequently treated with acid—some more successfully than the first.*

As a result of these exciting successes, and the proven use of corrosion inhibitor, interest in the industry was keen, spreading rapidly. From their Well Services Group, which handled brine wells, Dow Chemical fortuitously formed a new subsidiary, Dowell (from Dow Well Services) in November 1932.[10] Dowell is now a division of the giant French company, Schlumberger.

Several companies were quickly organized in 1932 to capitalize on the growing "carbonate acidizing" business.[11–16] These included two companies no longer in existence, the Oil Maker's Company and the Williams Brothers Treating Company. Another company was the Chemical Process Company, the predecessor to Byron Jackson, Inc., eventually BJ Services.[17]

In 1935 the Halliburton Oil Well Cementing Company decided to begin acidizing oil wells commercially. As for the operator, Pure Oil, that fine company was eventually merged into the Union Oil Company of California (now Unocal) in 1960, with its place and legacy in acidizing history unfortunately all but forgotten.

Since its first commercial use in 1932, hydrochloric acid has remained the primary acid for stimulating carbonate formations. For many years, 15% HCl was the standard mixture and has remained the most common. In 1961 Harris brought the use of acetic acid to the industry.[18] Because acetic acid is less corrosive than HCl, it was suggested that it could replace HCl in certain applications, particularly at high temperatures. Later, formic acid was also found to be useful in solving certain problems inherent to acidizing with HCl.

In 1966, Harris *et al.* delineated acid concentration effects and the practical aspects of using high concentrations of HCl.[19] Laboratory studies showed that the properties of solutions containing greater than 15% HCl had significantly different properties than lower concentration HCl solutions.

In many cases, the different properties of higher concentration HCl were beneficial to acidizing carbonates. It was found that deeper acid penetration could be achieved with the higher concentration solutions. Use of mixtures such as 20% HCl and 28% HCl became common. Lund *et al.* wrote benchmark papers describing the dissolution of dolomite and calcite in hydrochloric acid.[20, 21] These papers and others led to the current understanding and development of matrix acidizing fluid systems.

The development of acid fracturing did not come about until the late 1930s and early 1940s. Clason suggested that production increases observed in acidizing treatments in carbonates would be impossible from the radial penetration (matrix treatment) of acid.[22] He suggested that crevices (fractures) must be present, and only through the enlargement of the crevices and/or removal of drilling fluid or other deposits from the crevices or fractures could large increases in productivity be explained.

About 10 years later it was discovered that fractures could indeed be created. Fracturing theories soon developed. Since then, the development of fracturing technology, with and without propping agents, has been swift.

In 1972 Nierode and Williams presented a kinetic model for the reaction of hydrochloric acid with limestones.[23] This work raised carbonate acidizing from mystery to somewhat more predicable technology. Nierode and Williams used their model to predict acid reactions during fracturing operations and to design acid fracturing treatments. Work by many others, including that by Roberts and Guin, has led to ever-improving acid fracturing models used by service companies today.[24]

In the 1960s, treatment design and fluid development were emphasized. Since then, many acid fracturing systems and procedures have been developed to control fluid loss and retard reaction rates.

SANDSTONE ACIDIZING

Because of the growing excitement surrounding acid treatment of limestone formations throughout 1932, interest in developing treatments for sandstone formations began growing too. In May 1933, Halliburton conducted the first sandstone acidizing treatment using a mixture of hydrochloric and hydrofluoric acid (HF). The treatment was pumped in a test well belonging to the King Royalty Co., near Archer City, Texas. It was 1532 feet (ft) deep with 11 ft of open-hole production interval.[1] The exact composition and strength of the HF acid mixture is not known.

Unfortunately, the results of this first attempt were very discouraging. The reaction of the strong acid solution in the formation caused substantial sand production into the wellbore. Consequently, use of HCl-HF was not very popular for the next 20 years. In fact, Halliburton did not offer HCl-HF until the mid-1950s.[1]

Interestingly, Halliburton was apparently unaware of a patent application filed on March 16, 1933 by Jesse Russell Wilson of the Standard Oil Company of Indiana. This patent concerned the use of hydrofluoric acid for treating sandstone formations.[25] Another patent application was filed the same day by James G. Vandergrift, employing a mixture of mineral acid and hydrofluoric acid.[1]

The Wilson patent application indicated an understanding of sandstone acidizing and its purpose far ahead of its time. His description of the "problem," recounted in *Acidizing Fundamentals*, is as follows:

> *Finely divided sand, other siliceous and miscellaneous debris tend to be deposited by the fluid flowing toward the base of the well, thereby clogging up the pores or passages in the geological formation immediately surrounding the base of the well with the result that resistance to flow is greatly increased. It has occurred to me that one method of rectifying this situation is to dissolve out this deposited material by the use of a suitable reagent. In the case of sand, one suitable reagent is hydrofluoric acid or hydrogen fluoride which reacts with the sand . . . producing water and silicon tetrafluoride, the latter being a gas. The principal difficulty with this procedure is that hydrofluoric acid is an extremely dangerous material to handle. The risk encountered in introducing it into an oil well would be so great that I do not believe it has ever actually been attempted.*

Wilson's process called for generation of HF acid in the wellbore or in the formation. Looking back, his perception of acid-removable formation damage due to solids plugging is remarkable. Many sandstone acidizing treatments have been pumped since, without such understanding of purpose and potential.

Dowell did introduce a mixture of 12% HCl-3% HF, called "Mud Acid," in 1939. The purpose of this mud acid was to remove drilling mud filter cake from the wellbore. However, use was limited to wellbore treatment only. Successful treatments were pumped in the Gulf Coast area.[26, 27]

Despite the unpopularity, the early use of mud acid was still a major breakthrough in well stimulation technology. While HCl was found to be effective in removing carbonate deposits, it was clear that it was not adequate for dissolving clay and other siliceous deposits. The need for an acid capable of dissolving siliceous materials led to the use of HF acid for removal of drilling mud damage in sandstone formations. It took some time, but mud acid became the treatment of choice for mud-damaged sandstone formations. This 12% HCl-3% HF acid mixture is still quite common and is now known as "regular strength" mud acid.

The year 1965 marked another milestone in the development of sandstone acidizing technology. A paper by Smith and Hendrickson discussed the reactivity and kinetics of HF acid and the effects of common variables encountered in the field.[28] HF reactions with rock minerals and secondary depositions were studied theoretically. In addition, core flow tests were conducted with Berea sandstone cores.

This study removed much of the mystery in HF acidizing for petroleum engineers and led the way to improved design practices. The most important result of this work was the development of "tapered" HF treatments, with an HCl preflush and overflush, to inhibit deposition of plugging reaction products.

Numerous matrix acidizing treatments of sandstone formations have been conducted since the mid-1960s. In addition to HCl and various HCl-HF mixtures, acetic and formic acids have been used, alone and in combination with HF.

In the 1970s and early 1980s there was a proliferation of "novel" sandstone acidizing systems. Most were developed to provide certain benefits, such as:

1. Retarding HF spending to achieve deeper live acid penetration

2. Preventing precipitation of HF-rock reaction products

3. Preventing excessive acid reaction, which could lead to rock softening or compromised formation integrity

4. Stabilizing fine particles (clays, feldspars, quartz, etc.) that are able to migrate and can cause plugging in pore throats

For the most part, a properly designed conventional treatment with HCl-HF acid mixtures will stimulate damaged sandstone formations. Risks associated with acidizing, such as fines migration, precipitation of reaction products, and rock deconsolidation can normally be minimized with proper volumes and concentrations of acids used.

In the 1980s and into the 1990s, developments in sandstone acidizing addressed treatment execution more than fluid chemistry. Techniques included nitrified and foamed acid treatments, high-rate/high-volume HF acidizing, and CO_2 enhanced HF acidizing. These are discussed in later chapters.

Most recently, fluid chemistry has again stepped to the forefront. New or "novel" systems have appeared on the scene to combat problems such as reprecipitation of HF/rock reaction products, acid-sensitive formations, and minerals. New systems or, more often, twists on old systems, are developed primarily to set one service company apart from the next, or to respond to customers' perceived needs. In some cases, this does lead to advancements in acidizing treatment methodology and treatment results.

Acidizing history has shown that both success and failure are common, and both may be expected. However, failure and disappointment may lead to increased understanding and greater future success, if not in a particular field, certainly in fields and wells elsewhere. As a general treatment method, acidizing should never be written off because of disappointing results. As with all scientific advancement, great discovery and progress have been made in the field of acidizing when discouragement did not set in and overtake motivation and ingenuity.

REFERENCES

1. B. B. Williams, J. L. Gidley, and R. S. Schechter, *Acidizing Fundamentals,* monograph series (Dallas: Society of Petroleum Engineers, 1979).

2. "A Great Discovery," *Oil City Derrick* (Oct. 10, 1895).

3. S. W. Putnam, "Development of Acid Treatment of Oil Wells Involves Careful Study of Problems of Each," *Oil & Gas Journal* (Feb. 23, 1933): 8.

4. H. Frasch, "Increasing the Flow of Oil Wells," U.S. Patent No. 556,669 (March 17, 1896).

5. M. E. Chapman, "Some of the Theoretical and Practical Aspects of the Acid Treatment of Limestone Wells," *Oil & Gas Journal* (Oct. 12, 1933): 10.

6. *The Dow Chemical Company v Halliburton Oil Well Cementing Company,* Opinion of Circuit Court of Appeals, Sixth Circuit, U.S. Patent Quarterly 90.

7. "Acid Treatment Becomes Big Factor in Production," *Oil Weekly* (Oct. 10, 1932): 57.

8. R. B. Newcombe, "Acid Treatment for Increasing Oil Production," *Oil Weekly* (May 29, 1933): 19.

9. S. Putnam, "The Dowell Process to Increase Oil Production," *Industrial Engineering & Chemistry* (Feb. 20, 1933): 51.

10. "Chemical Company Forms Company to Treat Wells," *Oil Weekly* (Nov. 28, 1932): 59.

11. S. W. Putnam and W. A. Fry, "Chemically Controlled Acidation of Oil Wells," *Industrial Engineering & Chemistry* 26 (1934): 921.

12. "Chemical Treatment Halts Junking Breckenridge Wells," *Oil Weekly* (Feb. 13, 1932): 40.

13. "North Louisiana Operators Pleased with Acid Results," *Oil Weekly* (Dec. 19, 1932): 70.

14. C. E. Clason and J. G. Staudt, "Limestone Reservoir Rocks of Kansas React Favorably to Acid Treatment," *Oil & Gas Journal* (April 25, 1935): 53.

15. D. H. Bancroft, "Acid Tests Increase Production of Zwolle Wells," *Oil & Gas Journal* (Dec. 22, 1932): 42.

16. W. W. Moore, "Acid Treatment Proved Beneficial to North Louisiana Wells," *Oil Weekly* (Oct. 29, 1934): 31.

17. P. W. Pitzer and C. K. West, "Acid Treatment of Lime Wells Explained and Methods Described," *Oil & Gas Journal* (Nov. 22, 1934): 38.

18. O. E. Harris, "Applications of Acetic Acid to Well Completion, Stimulation and Reconditioning," *Journal of Petroleum Technology* (July 1961): 637.

19. O. E. Harris, A. R. Hendrickson, and A. W. Coulter, "High Concentration Acid Aids Stimulation Results in Carbonate Formations," *Journal of Petroleum Technology* (Oct. 1966): 1291.

20. K. Lund *et al.*, "Acidization I. The Dissolution of Dolomite in Hydrochloric Acid," *Chemical Engineering Science* 28 (1973): 691.

21. K. Lund, H. S. Fogler, and C. C. McCune, "Acidization II. The Dissolution of Calcite in Hydrochloric Acid," Chemical Engineering Science 30 (1975): 825.

22. C. E. Clason, "A New Conception of Acidizing," *The Petroleum Engineer* (1943).

23. D. E. Nierode, B. B. Williams, and C. C. Bombardieri, "Prediction of Stimulation from Acid Fracturing Treatments," *Journal of Canadian Petroleum Technology* (Oct.–Dec. 1972): 31.

24. L. D. Roberts and J. A. Guin, "A New Method for Predicting Acid Penetration Distance" (paper SPE 5155, presented at the Society of Petroleum Engineers 49th Annual Fall Meeting, Houston, TX, Oct. 6–9, 1974).

25. J. R. Wilson, "Well Treatment," U.S. Patent No. 1,990,969 (Feb. 12, 1935).

26. S. C. Morian, "Removal of Drilling Mud from Formation by Use of Acid," *Petroleum Engineering* (May 1940): 117.

27. H. L. Flood, "Current Developments in the Use of Acids and Other Chemicals in Oil Production Problems," *Petroleum Engineering* (Oct. 1940): 46.

28. C. F. Smith and A. R. Hendrickson, "Hydrofluoric Acid Stimulation of Sandstone Reservoirs," *Journal of Petroleum Technology* (Feb. 1965): 215.

2 Acid Treatment Categories

There are two general categories of acid treatments:

- Matrix acidizing
- Fracture acidizing

In matrix acidizing, the acid treatment is injected at matrix pressures, or below formation fracturing pressure. In fracture acidizing, all (or at least a significant portion) of the acid treatment is intentionally pumped above formation fracturing pressure.

MATRIX ACIDIZING

Matrix acidizing has application in both carbonate and sandstone formations. In sandstone formations, matrix acidizing treatments should be designed to remove or dissolve "acid-removable" damage or plugging in the perforations and in the formation pore network near the wellbore.

Theoretically, acid flows through the pore system, dissolving solids and fines entrained in pore throats and pore spaces that impede oil or gas flow.

Figure 2–1 depicts acid flow through a sandstone matrix pore system. As acid flows through pore channels, it is presumably able to dissolve small fines and particles present in pore spaces, pore throats, and along pore walls.

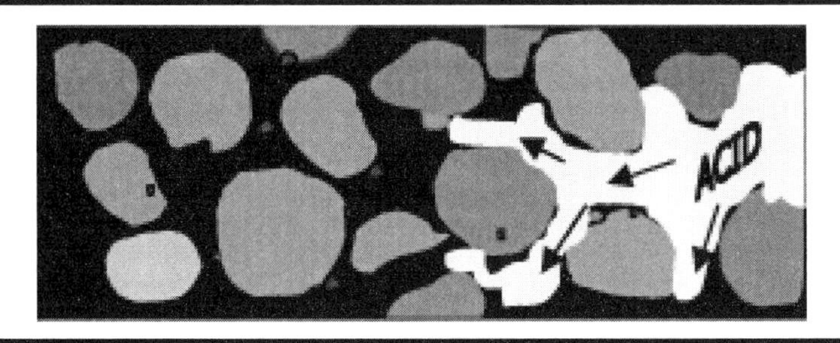

Fig. 2–1. *Acid flowing through a sandstone matrix pore system*

The majority of acid reaction is with pore-plugging or pore-lining solids and minerals. Therefore, in sandstone formations, matrix acidizing has application as a formation damage removal treatment. Generally, a sandstone acidizing treatment only has a chance for success if acid-removable plugging, or formation damage, is present. Matrix treatment of an undamaged formation cannot significantly increase production. There are certain exceptions, such as naturally fractured reservoirs. Sandstone acidizing will be discussed in part II.

In carbonate formations, matrix acidizing works by forming conductive channels, called "wormholes," through the formation rock. These penetrate beyond the near-wellbore region, or extend from perforations, as depicted in Figure 2–2. Acid-induced wormholes in carbonate rocks very much resemble the holes made by earthworms in the underground, hence the descriptor. Figure 2–2 is a simplification of conductive wormhole flow channels extending from perforations, with some branching (smaller channels "branching off" from main wormhole formed).

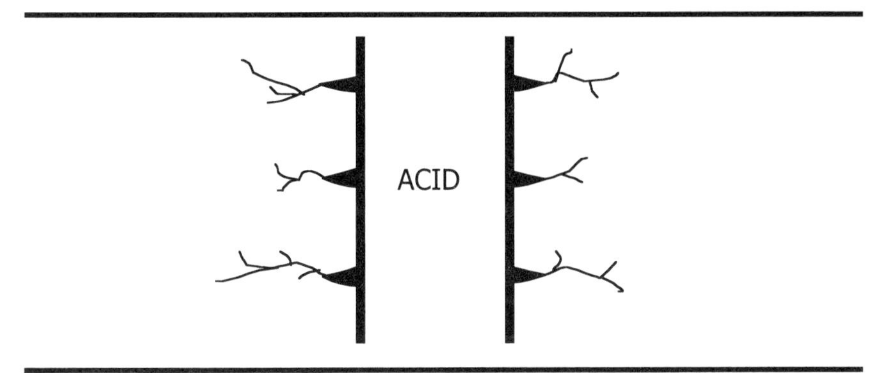

Fig. 2–2. *Acid-induced "wormholes"*

Quite often, acid will form predominately single wormholes from limited numbers of perforations, without significant branching. That is the case with strong acid, such as HCl. Weaker acids, such as acetic acid, and retarded acid systems tend to create more branching of wormholes, which is desirable, but only to a certain extent. Retarded acid systems include viscosified acids (e.g., gelled or emulsified acids). The nature of wormholes created depends on injection rate, temperature, and formation reaction characteristics as well.

In carbonate formations, matrix acidizing is principally a "damage bypass" treatment. If a carbonate formation is undamaged, a matrix acidizing treatment probably cannot be expected to do more than double the production rate. Carbonate matrix acidizing is discussed in part III of this book.

FRACTURE ACIDIZING

Fracture acidizing treatments are generally confined to carbonate formations. Acid fracturing treatments of carbonates are conducted either to bypass formation damage or to stimulate undamaged formations. This can include vugular and naturally fractured chalks, limestones, and dolomites.

Acid fracturing is an alternative to hydraulic fracturing with proppant. The objectives are the same: the creation of a long, open, conductive channel from the wellbore, extending deep into the formation. The basic principles of fracture propagation and geometry are also the same. The difference between the two fracturing methods is in how fracture conductivity is created and maintained.

Hydraulic fracturing uses proppant (e.g., sand) to hold the fracture open. Fracture acidizing does not use proppant. Fracture acidizing relies on the etching of fracture faces with acid to provide the required conductivity. Acid is injected into a fracture created by a viscous fluid (pad) or is itself used to create the fracture. As acid travels down the fracture, acid is transported to the fracture walls, resulting in dissolution etching.

If etching is nonuniform (differential), then the fracture may close with conductivity retained, as there will be low and high spots—voids and points of support—holding the channel open. Figure 2–3 shows two halves of a long core opened to show the rock surface dissolution resulting from an experiment in which acid was injected, simulating fracture acidizing.

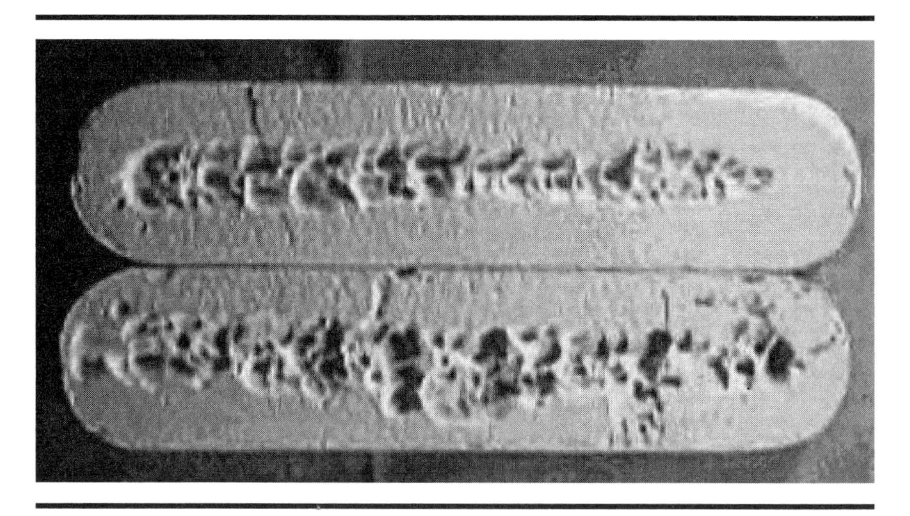

Fig. 2–3. *Two halves of long core sample showing acid dissolution (courtesy STIM-LAB, Inc., Duncan, Oklahoma; Internet website http://www.stimlab.com)*

The dissolution patterns show high spots and troughs. They form a conductive channel when the two halves are placed together—as long as the open channel can be supported under closure at the contact points. Fracture acidizing is discussed in part III.

ACIDS USED

In carbonate acidizing, acids commonly used are:

- Hydrochloric (HCl)
- Acetic (CH_3COOH)
- Formic (HCOOH)

HCl is the most common acid used. Organic acids, acetic and formic, came into use because they are less corrosive than HCl. Therefore, their primary benefit is for high-temperature applications.

Formic acid is more strongly reacting than acetic and is closer to HCl in strength, albeit weaker. Therefore, formic acid is not often used alone, although it can be. When it is necessary to replace HCl completely with a weaker acid, acetic is the common and most logical choice. Acetic and formic acids are more commonly used in combination with HCl. HCl-acetic, HCl-formic, and formic-acetic blends exist for high-temperature acidizing applications.

The formic-acetic blends are used to a lesser extent. They have been successfully applied in high-temperature fracture acidizing applications, where slower reaction and greater depth of live acid penetration into the fracture are required.

In rare cases, other acids may be used. Examples are sulfamic acid and chlorinated acetic acids. These acids are solids. Their use is limited mostly to remote location applications, where transporting liquid is impractical or prohibitively expensive. Sulfamic acid is a weak acid that has very limited application in cleanup treatments. Sulfamic acid decomposes at around 180 °F. Sulfamic acid comes in "stick" form. These acid sticks can be dropped into a wellbore, where they dissolve in water downhole to generate acid in solution. Chloroacetic acid is also a solid, so it can be used in remote locations where liquid transport is not possible. It is a fairly strong acid, as a result of the chlorination and the tendency to higher ionization. Chloroacetic acid is stronger than the organic acids (acetic and formic) but weaker than HCl.

In sandstone acidizing, acids commonly used are:

- Hydrochloric (HCl)
- Acetic (CH_3COOH)
- Formic (HCOOH)
- Hydrofluoric (HF)

HF is used most commonly in combination with HCl. It should never be pumped alone. It may also be used in combination with the organic acids, acetic and formic, or in combination with "acid blends," such as acetic-formic, HCl-acetic, and HCl-formic. Other organic acids, such as citric acid or proprietary organic acid systems, also may be combined with HF in matrix acid treatments of sandstone formations. Appendix A gives examples of successful sandstone acid treatment procedures.

Before delving into acid treatment design, chapter 3 follows with a general discussion of formation damage. The assessment of formation damage is the most important aspect of acid treatment candidate selection and treatment design development.

3 Formation Damage

Much has been written on the subject of formation damage over the years. For background, the best place to start is with the landmark paper on formation damage by Krueger.[1] Another somewhat forgotten but very important work is that by Maly.[2] For our purposes, this chapter will concern itself with formation damage issues as they relate to acid treatment design.

It is in the removal of near-wellbore formation damage that acidizing finds its primary application. In a well producing in radial flow regime, most of the pressure drop to the wellbore occurs within a short penetration distance into the reservoir. In fact, 50% of the total flowing pressure gradient takes place within 20 ft of the wellbore. If no damage is present, 25% of the pressure gradient is within 1–3 ft of the wellbore. If formation damage is present, it will contribute largely to pressure drop and will dominate well performance.

With respect to acidizing, especially sandstone acidizing, assessment of formation damage is perhaps the single most important factor in treatment design. Therefore, the design process for acidizing begins with candidate selection and formation damage assessment.

To assess formation damage, it is first necessary to understand the skin term in the Darcy's Law equation defining well production rate, and its effect on production rate. Well production rate, defined by the simple form of Darcy's Law for steady-state liquid flow in a radial reservoir, is as follows:

$$q = 7.082kh\,(p_e - p_{wf}) \,/\, [B\mu\,(\ln r_e/r_w) + s]$$

where

q	is production rate, barrels per day (b/d)
k	is the permeability, millidarcies (mD)
h	is the formation height, ft
p_e	is the reservoir pressure, psi
p_{wf}	is the flowing wellbore pressure, psi
B	is the formation volume factor, reservoir vol./prod. vol. (RB/STB)
μ	is the formation fluid viscosity, centipoise (cp)
r_e	is the reservoir radius, ft
r_w	is the wellbore radius, ft
s	is the skin value

It is important to note that production rate, q, is directly proportional to permeability, and inversely proportional to skin, s. Along with reservoir quality, these two variables, k (permeability) and s (skin), are of greatest importance in stimulation design. The combination of low permeability and high skin make for a very unproductive well. These two factors for a well must be understood and defined before considering a well stimulation procedure.

Skin damage is a mathematical representation of the degree of damage present. It can be represented, qualitatively, by the Hawkins equation, as follows:[3]

$$s = (k/k_s - 1) \times \ln(r_s/r_w)$$

where

k	is the formation permeability
k_s	is the permeability of altered zone extending to radius r_s
r_s	is the radius of altered (damaged) zone
r_w	is the wellbore radius

If a well is damaged ($k_s < k$), s will be positive. The greater the contrast between k_s and k, and the deeper the formation damage (r_s), the larger the value of s. Skin values can be quite large in extremely damaged cases. Skin approaches infinity in a totally damaged case ($k_s = 0$). If a well is stimulated ($k_s > k$), s will be negative.

The deeper the stimulation, the lower the value of s. However, skin values less than –5, or so, are rare. Such negative skins exist only in wells with deep, conductive hydraulic fractures (propped). Slightly negative skins may also arise in frac-and-pack completions, and in naturally fractured formations. If a well is neither damaged nor stimulated, then $s = 0$ (undamaged).

Permeability and skin can be measured with a pressure transient well test.[3] One should be conducted, whenever possible, both before and after stimulation. From a well test, measured skin, s, is really a multicomponent composite or total skin (s_t) term, comprised of different skin "contributors," as follows:

$$s_t = s_{c+\phi} + s_p + s_d + \Sigma_{pskins}$$

where

$s_{c+\phi}$	is the skin due to partial completion
s_p	is the skin due to incomplete perforations

s_d	is the skin due to damage
Σ_{pskins}	is the sum of pseudo-skin factors (phase- and rate-dependent effects)

Wells that show high skin values measured from well testing (upwards of 30, 50, 100 or even 300 or more) may be expected to have skin from other "nondamage" effects. In such cases, stimulation can still be substantial, as long as s_d is significant.

Sandstone acidizing is a method for removing acid-removable s_d only. Other than in very rare cases, production rate from an undamaged well producing from a sandstone formation could be increased up to perhaps 2 times the original rate, at best. Acid-removable skin, discussed in chapter 6, may manifest itself in the wellbore, the perforations, or within the formation.

Carbonate acidizing is, for the most part, a method for bypassing skin damage, rather than removing it directly, as is the case in sandstone acidizing. In addition, unlike an acid treatment in an undamaged sandstone formation ($s_d = 0$), an acid treatment in an undamaged carbonate formation can result in sufficient stimulation in certain cases. In particular, this is true with acid fracturing. This is because fracturing can stimulate a formation by effectively extending the wellbore radius, independent of whether skin damage is present.

However, greater degree of stimulation is achievable if damage is present to start with. Also, matrix acidizing in an undamaged carbonate formation can increase productivity ratio to a greater extent than in acidizing an undamaged sandstone formation. Still, as mentioned in the last chapter, production rate from an undamaged well producing from a carbonate formation may only be doubled. (Production rate may be somewhat more with a deeply penetrating, well-placed matrix acid treatment.)

Carbonate matrix acidizing is, realistically, a method for bypassing damage. Stimulation response potential is much greater when damage is present that can be bypassed by flow channels (wormholes) created by formation dissolution by acid.

Formation damage can occur during any well operation, including:

- Drilling
- Cementing
- Perforating
- Production
- Workover
- Stimulation

There is no well operation that is truly nondamaging. Any invasive operation may be damaging to well productivity. Production itself can cause damage. Therefore, in assessing formation damage, all aspects of a well and its history should be investigated, including:

1. Reservoir geology and mineralogy
2. Reservoir fluids
3. Offset well production
4. Production history
5. Drilling history (including fluids used)
6. Cementing program (including cement bond logs)
7. Completion and perforation reports (including fluids used)
8. Workover history
9. Stimulation history

RESERVOIR GEOLOGY AND MINERALOGY

In assessing damage, it is necessary first to find information on the reservoir geology as well as the mineralogy. An understanding of the rock type (sandstone or carbonate) and other features, including permeability and porosity, is most important. This includes the nature of porosity (matrix vs. naturally fractured) and how permeability is distributed (how it varies) across the producing interval, or injection interval, as the case may be.

I was once asked to design a matrix acidizing treatment for an exploratory well in the Middle East. The acid treatment was expected to be required to induce production and continue with the exploration program. My first question was, "What type of formation is it?" The answer was, "It is either a carbonate or a sandstone." I then contacted the service company representative assigned to this treatment to ask if he knew more.

When I told the representative that my understanding was that the formation was either a carbonate or a sandstone, he replied that that was more than he had been told. Sometimes there is not enough information to go on to design a treatment. That did not stop us, though, as we had no choice. The point is, whatever information is known about the reservoir must be gathered and applied.

The mineralogical characteristics are also important. This includes knowledge of the bulk mineralogy, which defines the types of minerals present. It also takes into account the location of mineral phases in and around the rock pore spaces, or natural fracture network, if present.

For example, in a particular sandstone, there may be 10% clay present; however, all is lining the walls of pore spaces. If acid were to be injected into this matrix, it would see a much higher percentage of clay relative to other mineral phases. These other mineral phases might include large quartz framework grains, which may represent 80% of the bulk mineralogy. In such a case, an acid treatment more mild than would otherwise be considered may be needed.

An example with a carbonate formation may be one in which 95% is calcium carbonate ($CaCO_3$). The remaining 5% represents "other" sandstone minerals partially filling or lining natural fractures, the principal flow channels in need of stimulation. Acid injection in such a formation would preferentially enter natural fractures and initially would contact predominately the noncarbonate mineral phase. This could result in inadequate stimulation, or release of siliceous particles and subsequent plugging. This would reduce natural fracture conductivity, rather than increase it.

Gray and Rex provide useful references on clays.[4] Ali is a good resource for sandstone mineralogy.[5]

RESERVOIR FLUIDS

Next, knowledge of the reservoir fluids is essential. This includes the fluid types (oil or gas) and fluid properties. Fluid properties include CO_2 or H_2S content in gas, gravity of the oil, paraffin and asphaltene contents, and produced water volume and properties (ionic composition and scaling tendency).

Reservoir fluid sample analyses should be reviewed, or conducted if they do not exist. Produced fluids can cause damage through deposition of wax or asphaltenes from oil, or from scale formed from produced brine. Also, as well assessment continues, it may be found that fluids incompatible with the reservoir fluids may have been introduced sometime in its past, inducing organic deposition or scale formation.

OFFSET WELL PRODUCTION

Offset and even nearby well comparison is a first-step indicator that a well of interest in an existing field is underperforming and might be a candidate for stimulation. A well with good bottomhole pressure, or at least comparable to offset wells, but with lower production, may very well be suffering from some form of damage.

PRODUCTION HISTORY

The production history for an older well, or at least one that has been producing for some time, must be reviewed carefully and initially in the well assessment process. A sudden and sharp decline in production is the strongest indicator of damage. Often it is indicative of migration of mobile formation fines in the near-wellbore region. Similarly, in a new well, a good DST (drill stem test) but poor performance after completion indicate damage in the completion process.

There are several possible production damage mechanisms, depending on the type of well (oil or gas) and formation characteristics. Such damage contributors include:

- Fines migration
- Inorganic scale deposition
- Organic solids deposition (such as paraffin and asphaltenes)

FINES MIGRATION

Fines migration can occur in sandstones during abrupt increases in production, including those following workover or stimulation. It may occur slowly during the natural production flow of the well too, if production rates exceed a flow velocity above which fines are released. Such critical velocities exist in both oil and gas wells. Core test evaluation can provide some handle on flow-induced fines migration. Fines that are mobile in producing fluids bridge and accumulate at pore throats and constrictions, thereby reducing matrix permeability.[1, 6, 7]

Fines migration may be induced and exacerbated in sandstones by acidizing, especially by those treatments that use HF acid. Although often designed to remove fines, HF treatments are notorious for generating new fines or releasing existing, undissolved fines. These are principally from the reaction of HF with siliceous mineral fines present within pore spaces (such as clays, feldspars, quartz fines, etc.). Fines migration during production following an acid treatment can be especially severe, and more rapidly damaging than prior to stimulation treatment.

INORGANIC SCALE DEPOSITION

Deposition of inorganic scale may occur during well production. Depending on well conditions and produced water characteristics, different scale types may form. Often, scale formation is associated with a breakthrough of water production. Common scales include calcium carbonate, iron carbonate, calcium sulfate, barium sulfate, strontium sulfate, and iron sulfide. Combinations may also form.

Unfortunately, not all scales are removable by chemical methods. For example, only calcium carbonate scale is readily removed by acid. Others, such as calcium sulfate, can be removed chemically, but inefficiently or only partially. Most require mechanical methods, such as milling, for complete removal. See chapter 14 for further discussion.

Organic deposition

Downhole organic deposition is a common oil well problem. If not properly diagnosed, it can be missed or mistaken for other forms of damage. There are two general types of organic deposition products in oil wells: paraffin (wax) and asphaltene. Paraffin may also occur in gas wells producing condensate. Neither is soluble in acid. They must be treated with solvent, preferably an aromatic, such as xylene.

Paraffin. Paraffin deposition is a function of reservoir or wellbore temperature. If the temperature is above the cloud point (the point at which the paraffin present in the oil begins to deposit), deposition will not take place. However, a slight reduction in temperature can cause the paraffin to crystallize (clouding the oil) and deposit in perforations or in the wellbore. Such a temperature change may occur during any well operation in which fluid is introduced from the surface to the wellbore or formation.

Downhole temperature will be temporarily reduced when foreign fluids contact the formation oil. Once solid paraffin has begun to form, in order to resolubilize, an increase in the oil temperature above the well temperature is usually required. Unfortunately, this is often difficult, if not impossible, to accomplish. Repeated treatment with solvent may be necessary.

Asphaltenes. Asphaltenes are high carbon number, primarily cyclic hydrocarbons present in crude oil in colloidal suspension. Asphaltenes may drop out of solution as a very nasty, persistent, damaging deposit downhole.

Unlike paraffin deposition, asphaltene deposition is not sensitive to temperature, but to pressure drop. Localized pressure drops or disturbances occur as crude oil flows from the formation into the wellbore and into the production tubing can cause asphaltenes to begin depositing. Upsets to the fluid equilibria can also cause deposition of asphaltenes. This can occur any time fluid is injected into the formation, especially those with extreme pH. Examples of such fluids are acid (low pH), cement filtrate (high pH), drilling mud (high pH), and certain completion fluids (low or high pH).

It is especially important to identify organic deposition in an acid treatment candidate well, because this damage is not acid-removable. In fact, contacting organic deposits with acid can create more severe (perhaps irreversible) damage. In addition, assessment and identification of organic deposits must be made so that proper pretreatment can be incorporated into the stimulation design.

If at all possible, downhole samples should be collected for analysis of paraffin and asphaltene and their potential for deposition. Acid/crude oil compatibility tests should be conducted by the service company to design a compatible acid system. Removal of organic deposits is discussed in chapter 15.

DRILLING HISTORY

Review of the drilling history (drilling reports) can reveal damage caused early in the life of a well. This includes an understanding of fluids used and their properties (oil-based or water-based, pH, solids used, etc.). There are two primary drilling damage mechanisms to consider:[8–10]

- Drilling mud filtrate loss to the formation
- Drill solids invasion

Drilling practices, in general, have become less damaging in recent years, especially with the advent of oil-based systems and lower solids requirements. Decreased use of fresh water drilling muds in sandstones has largely contributed to less damaging practices, as well.

However, to say that a drilling operation is nondamaging to the formation will always be wrong. Certainly, wells are drilled without causing formation damage. However, no drilling method is inherently nondamaging. Therefore, the drilling process should always be viewed as potentially damaging in any case.

MUD FILTRATE LOSS TO THE FORMATION

Damage from mud filtrate loss to the formation includes permeability reducing effects, such as:

- Invasion of unbroken viscosified gel fluid filter cake on the formation face or within natural fractures or vugs (void pore spaces)

- Alteration of wettability from water-wet to the undesirable oil-wet state caused by surfactants, or oil-based fluids
- Reaction of low salinity or high-pH fluid filtrate (pH ~12) with formation minerals causing swelling or dispersion of sensitive clays, and migration of fines through clay effects or mineral dissolution[11–15]

Drill solids invasion

Damage caused by drill solids invasion includes:

- Plugging and bridging by weighting material, such as bentonite clay (partially soluble in HF mixtures) or barite (barium sulfate, $BaSO_4$, which is insoluble in acid)
- Loss of and subsequent plugging by drill cuttings and cuttings fines
- Loss circulation material (LCM) entering the formation vugs or natural fractures
- Pipe dope and other miscellaneous solid materials used

Whole mud loss to formation

In naturally fractured formations, an additional damage mechanism is the loss of whole mud deep into the formation. This form of damage is very difficult to remove entirely. Such damage may require hydraulic fracturing or repeated sequences of stimulation fluid injection and production, which is not always possible.

CEMENTING PROGRAM

Damage during cementing probably does not occur as often as in other operations. However, the cementing program and report must still be investigated for the following possibilities:

- High losses of high-pH cement filtrate; high-pH filtrate can disturb clays and induce fines migration, particularly in sandstones[16]

- Invasion and plugging by cement solids
- Loss of whole cement to the formation, either into natural fractures or during inadvertent fracturing during cementing

COMPLETION AND PERFORATION REPORTS

Completion and perforation history must then be evaluated. Completion practices, and the fluids used, are potentially damaging. For example, perforating in dirty fluids can be very damaging. Unfiltered solids in perforating fluid are injected at such high velocity during the perforation process, that plugging can be extremely severe.

Perforating in oil-based drilling fluid also may cause undesirable wettability alteration, reducing relative permeability to oil or gas. This is not often considered a problem, because oil-based fluids are often erroneously considered nondamaging. Wettability alteration may often be treated with a solvent/water-wetting agent. However, it is not always a simple matter to do so.

A common damage mechanism is compaction or crushing of the formation within a thin rim around the perforation created. In extreme cases, the altered permeability region surrounding the perforation can be a small fraction of the native permeability. This can drastically reduce the inflow of fluid into the perforation.

The presence of this type of damage is difficult to assess. Fortunately, acidizing can usually remove or bypass such damage. Perforating debris may be a problem, as well. There is always debris generated by the perforation process. This can be especially serious in gravel pack completions, if debris is not cleaned from perforations before placing the pack. Debris is not always acid-removable.

Completion fluids are not always compatible with formation mineralogy. This is of more concern in water-sensitive sandstones than in carbonates. Core flow tests can be used to evaluate sensitivity of the formation to various brine formulations. Core testing can be used to select the proper fluid or to modify a fluid with additives to address water-sensitivity.

WORKOVER HISTORY

Damage is often created during workover operations. Workover history must be investigated thoroughly to uncover one or more possible damage contributors, including:

- Use of dirty fluid, which can cause plugging
- Use of a workover or kill fluid that is incompatible with formation brine (e.g., seawater), which can result in the formation of a variety of scales: carbonates (acid-soluble) and sulfates (acid-insoluble)
- Paraffin deposition resulting from near-wellbore fluid temperature reduction
- Water-blocking (retention of water in formation pore spaces)

The use of seawater, in particular, may cause downhole scaling when contacting incompatible formation brines. Simple laboratory testing with fluid samples can be run to evaluate fluid compatibility. The nature of the workover itself may be physically damaging to the formation.

STIMULATION HISTORY

Review of stimulation history is of the utmost importance. It should not be avoided, no matter what one may be in fear of discovering. Valuable information and even correlations may be extracted; however, not at the expense of taking documented past stimulation results too far. There are a number of circumstances and well conditions that may be factors in poor stimulation responses that cannot be discovered or explained.

In any case, stimulation history must be investigated to uncover possible damage caused, as well as for guidance in subsequent stimulation design. Unfortunately, records of past stimulation treatments are often incomplete or sketchy. They are not often written with future stimulation needs in mind.

In reviewing stimulation history, one should contact those who may have been involved in past treatment design, or were on-site during treatment, provided they are still available and their memories are intact. Little things that were observed may shed light on what may have happened in a past treatment.

The key point is that stimulation can very easily cause damage rather than remove or bypass damage successfully. Damage can occur during acidizing and fracturing, in both sandstones and carbonates.

ACIDIZING DAMAGE MECHANISMS

Acidizing damage mechanisms include:

- Inadvertent injection of solids
- Use of incompatible additives or improper mixing procedures
- Reprecipitation of acid reaction products
- Loss of near-wellbore formation compressive strength
- Formation of emulsions
- Formation of sludge
- Water blocking
- Wettability alteration
- Unbroken gel plugging (carbonate acidizing)
- Posttreatment fines migration

INADVERTENT INJECTION OF SOLIDS

Injection of solids can occur if dirty fluids are used. Also, if acid is pumped through dirty tubing, or tubing containing rust or mill scale (new tubing), iron-containing solids can be injected. Fortunately, incorporating an acid pickling pretreatment stage in an acid job can largely prevent this (see chapter 6).

INCOMPATIBLE ACIDS/IMPROPER MIXING

Use of acid additives that are incompatible with one or more other additives, or incompatible with the formation or formation fluids, can cause damage, sometimes irreversibly. This problem can be eliminated with proper additive selection (see chapter 6) and quality control practices (see chapter 16).

Reprecipitation of reaction products

The reprecipitation of reaction products is not a serious concern in carbonate acidizing, but it is in sandstone acidizing. Gdanski and Peavy, and Shuchart, among others, have conducted analyses of fluid returns from HF acid treatments.[17, 18] Their research indicates that many reactions take place in the formation as HF injection proceeds. Coulter and Jennings refer to the work of Gdanski and Peavy, and others, and summarize HF reactions and associated problems, as follows:[19]

> *This work, reported in several technical papers, provides insight into many of the problems associated with HF acidizing over the years. The work reported cites primary through tertiary reactions taking place when pumping HF fluids on sandstone formations containing alumino-silicates. The primary reaction is that of the HF on alumino-silicates resulting in silicon fluorides. These reaction products can continue to react on alumino-silicates, a secondary reaction, at temperatures above ~ 100 °F. During this secondary reaction, clay material is brought into solution and any sodium and potassium ions present can react with any unreacted silicon fluorides causing a precipitate of sodium or potassium fluosilicate. Additionally, as the reaction continues with all the silicon fluoride reacted, aluminum fluoride complexes are formed which react on alumino-silicates, a tertiary reaction. At this point, as the HCl is consumed or the reaction products are diluted with formation water and the pH elevated, an aluminum fluoride precipitate can be formed creating damage.*

If that sounds complicated, it is because it is. Besides that, dissolved iron will precipitate as iron compounds when acid spends. (Oxides of Fe^{3+} precipitate at pH > ~2, therefore, they present a problem; oxides of Fe^{2+} precipitate at pH > ~7, therefore they do not present a serious problem.) Also, HF will react with calcium and magnesium ions to form insoluble calcium and magnesium fluoride salts (CaF_2, MgF_2).

I never seem to be able to fully grasp the entire story of HF reprecipitation reactions and their consequences.[20, 21] Fortunately, considerable work has been conducted to shed some light in this area.[22–28] There are numerous potentially damaging HF reaction products. The damaging effects of these reactions and their relative importance are subject to controversy. Certain problems may exist or become of greater concern in order to develop and market novel acid additives and acid systems.

However, one should be aware that there are many reprecipitation reactions that do legitimately take place to the extent that they should be of concern. Reprecipitation problems are most likely and severe when HF acid treatments are shut in near the wellbore for more than several hours before producing the well back. Table 3–1 summarizes damaging HF reactions and precipitates.

While the reprecipitation reactions in HF acidizing are various and complex, they do not preclude success. Reprecipitation reactions are unavoidable, but their effect on stimulation response can be minimized with proper fluid selection and treatment design, as discussed in chapter 6. So, there is hope and reason to be encouraged.

Loss of Near-Wellbore Compressive Strength

Loss of near-wellbore compressive strength is another acidizing-related damage mechanism that may occur in both carbonates and sandstones. In sandstones, excessive volume of HF acid, or use of HF in a concentration that is too high for the formation, may weaken the near-wellbore matrix. This can result in fines migration or, even worse, sand production. Perforations may slough or "collapse," as well, thereby restricting flow capacity to the wellbore. Acid may also remove the natural cementation holding quartz grains together.

Formation cementation may be carbonate mineral or clay. If a sandstone matrix is held together primarily by carbonate, acidizing may substantially remove this mineral phase, resulting in sand production and severely impaired production. Expensive workover operations are likely in such a case.

Reaction	Precipitate(s)
HF + carbonates (calcite, dolomite)	Calcium and magnesium fluoride (CaF_2, MgF_2)
HF + clays, silicates	Amorphous silica (orthosilicic acid, H_4SiOH_4)
HF + feldspars	Sodium and potassium fluosilicates (Na_2SiF_6, K_2SiF_6)
HF + clays, feldspars	Aluminum fluorides (AlF_n^{3-n}) Aluminum hydroxides
HF + illite clay	Na_2SiF_6, K_2SiF_6
Spent HF + formation brine, seawater	Na_2SiF_6, K_2SiF_6
HCl-HF + iron oxides and iron minerals	Iron compounds
HF + calcite (calcium carbonate)	Calcium fluosilicate

Table 3–1. *Damaging HF Reactions in Sandstones*

Excessive acidization of a carbonate, especially in matrix acidizing, may create large voids or caverns extending from the wellbore into the formation. These voids may collapse in on themselves, resulting in poorer productivity from perforations or in open-hole conditions. Again, the problems associated with excessive reaction can be reduced or avoided with proper treatment design.

FORMATION OF EMULSIONS AND SLUDGES

Emulsions and sludges can form as a consequence of incompatibilities between the combinations of acid, formation fluids, and acid reaction products. Proper planning, additive selection, and fluid compatibility testing will prevent these damage problems to a great extent.

WATER-BLOCKING AND WETTABILITY ALTERATION

Water-blocking and wettability alteration are somewhat related acidizing damage mechanisms, especially in sandstones. In tight formations, water that is introduced to the formation may be retained by capillary forces. Gas or oil production rates may be severely impaired. Proper selection of surfactant additives is necessary to avoid water-blocking. This is quite often difficult to assess without the benefit of core flow testing with representative formation core.

However, in most cases, proper additive selection will greatly reduce the potential for water blocking. Water blocks may be removed with subsequent treatment with an aromatic solvent containing mutual solvent and surfactant. Solvent may be an aromatic hydrocarbon, such as xylene, or water. Mutual solvent could be EGMBE (ethyleneglycolmonobutylether) or the like. Aqueous-based treatments exist, as well. Water, or brine, containing a surfactant and mutual solvent or alcohol, may also be effective.

UNBROKEN GEL PLUGGING

Unbroken gel plugging may occur in carbonate acidizing procedures in which gelled acid is used. Monitoring and analysis of posttreatment flowback fluid samples may indicate unbroken gel, which will restrict flow of formation fluids. Gel breaking should be defined and controlled through pretreatment laboratory testing by the service company, at expected downhole temperature and residence time conditions.

POST-TREATMENT FINES MIGRATION

Posttreatment fines migration is quite common in sandstone acidizing. It may be difficult to avoid in many cases. The reaction of HF with clays and other aluminosilicate minerals, and quartz, can release undissolved fines. Also, new fines may be generated as a result of partial reaction with high-surface-area minerals, particularly the clays.

Fines generated by the HF reaction create a two-fold problem. First, the plugging at pore throats by these mobile fines reduces permeability. In addition, fines generated may also stabilize emulsions formed by the interaction of acid, or spending acid, with crude oil. This could occur in both sandstones and carbonates containing siliceous fines. In carbonate acid fracturing operations, it is also possible to release noncarbonate fines. These fines can then fill or bridge in low conductivity acid-etched fractures created, or at "tight" spots in the fracture.

Postacidizing fines migration problems can be reduced by bringing a well on slowly after acidizing, and increasing rate step-wise over time (e.g., one to two weeks), rather than maximizing return production right away. In this way, fines are given less opportunity to collect at pore throats or flow restrictions all at once, jamming flow paths. It is a bottleneck, or traffic engineering problem, really. Well flowback should be treated in similar manner. Recommended flowback practices are discussed further in chapter 16.

HYDRAULIC FRACTURING DAMAGE MECHANISMS

Damage can be caused during hydraulic fracturing operations too. Significant fracturing damage mechanisms include:

- Fines migration and plugging in the proppant pack
- Problems associated with incompatible fluids
- Unbroken polymer gel in the propped fracture

FINES MIGRATION AND PLUGGING IN THE PROPPANT PACK

Fines migration and plugging in the proppant pack may occur from proppant crushing at high fracture closure pressure. This is possible if a proppant with inadequate strength is used, or if high drawdown is applied to the fracture proppant pack. Formation fines may produce into the proppant pack, causing plugging in much the same way they might damage a gravel pack.

Fines migration following fracturing is not as common a problem as it is in matrix production flow (radial flow). Hydraulic fracturing can often reduce the tendency for fines to mobilize in a formation susceptible to fines migration, as production through the fracture is dominantly linear flow, and pressure drawdown that induces fines migration is largely reduced.

INCOMPATIBLE FRACTURE FLUID ADDITIVES

Incompatible fracture fluid additives may cause damage in much the same way as with acidizing fluids. In fracturing, the greater concerns would be with the potential for water-blocking in the matrix feeding the fracture created, or with emulsion formation. Again, these may be avoided with pretreatment fluid compatibility testing.

UNBROKEN GEL

Unbroken gel will cause damage. Most fracturing fluids are designed to break viscosity on their own, or by inducement with specific gel breaker additives. Gel must break down in viscosity to a thin fluid that easily flows back following fracturing stimulation treatment. Gel plugging may result if improper or inadequate breaker is used.

This type of damage may be removed with acid treatment (HCl) or with special polymer-specific enzyme treatments, available from certain service companies. Sometimes, in high-temperature applications, if polymer remains in the formation for an extended period of time, the polymer continues to thicken or "congeal." This can result in a plug that is more difficult to remove.

There are many potential damage mechanisms and types of formation damage that may exist in a well. Some are acid-removable. Some are not. Those types of damage that are removable with acid may require a specific acid type with certain additives. It is therefore most important to assess and understand as best as possible if damage is present, and if so, the damage type. Chapter 6 further discusses acid-removable formation damage.

REFERENCES

1. R. F. Krueger, "An Overview of Formation Damage and Well Productivity in Oilfield Operations," *Journal of Petroleum Technology* (Feb. 1986): 131–52; *Transactions*, American Institute of Mechanical Engineers, 281.

2. G. P. Maly, "Close Attention to the Smallest Details Vital for Minimizing Formation Damage" (proceedings, Society of Petroleum Engineers Symposium on Formation Damage Control, Houston, TX, Feb. 1976), 127–46.

3. M.F. Hawkins, "A Note on the Skin Effect", Journal of Petroleum Technology (Dec. 1956): 356–57.

4. D. H. Gray and R. W. Rex, *Clays and Clay Minerals* (Elmsford, NY: Pergamon Press, 1966): 355–66.

5. *Sandstone Diagenesis: Applications to Hydrocarbon Exploration and Production,* compiled by Syed A. Ali (Gulf Science and Technology Company, Dec. 1981).

6. T. W. Muecke, "Formation Fines and Factors Controlling Their Movement in Porous Media," *Journal of Petroleum Technology* (Feb. 1979): 144–52.

7. C. Gruesbeck and R. E. Collins, "Entrainment and Deposition of Fine Particles in Porous Media," *Society of Petroleum Engineers Journal* (Dec. 1982): 847—56 (paper SPE 8430) .

8. T. J. Nowak and R. F. Krueger, "The Effect of Mud Filtrates and Mud Particles Upon the Permeability of Cores," *Drilling and Production Practices* (American Petroleum Institute, 1951): 164–81.

9. E. E. Glenn and M. L. Slusser, "Factors Affecting Well Productivity—II. Drilling Fluid Particle Invasion Into Porous Media," *Transactions*, American Institute of Mechanical Engineers (1957), 210: 132–39.

10. A. Abrams, "Mud Design To Minimize Rock Impairment Due to Particle Invasion," *Journal of Petroleum Technology* (May 1977): 586–92.

11. P. H. Monaghan *et al.*, "Laboratory Studies of Formation Damage in Sands Containing Clays," *Transactions*, American Institute of Mechanical Engineers (1959), 216:209–25.

12. F. O. Jones, "Influence of Chemical Composition of Water on Clay Blocking of Permeability," *Journal of Petroleum Technology* (April 1964): 441–46; *Transactions*, American Institute of Mechanical Engineers, 231.

13. N. Mungan, "Permeability Reduction Through Changes in pH and Salinity," *Journal of Petroleum Technology* (Dec. 1965): 1449–53; *Transactions*, American Institute of Mechanical Engineers, 234.

14. S. F. Kia, H. S. Fogler, and M. G. Reed, "Effect of pH on Colloidally Induced Fines Migration," *Journal of Colloid and Interface Science*, 118, no. 1 (July 1987): 158–68.

15. R. N. Vaidya and H. S. Fogler, "Fines Migration and Formation Damage: Influence of pH and Ion Exchange" (paper SPE 19413, presented at the Society of Petroleum Engineers Formation Damage Control Symposium, Lafayette, LA, Feb. 22–23, 1990), 125–132.

16. W. C. Cunningham and D. K. Smith, "Effect of Salt Cement Filtrate on Subsurface Formations," *Journal of Petroleum Technology* (March 1968): 259–64.

17. R. D. Gdanski and M. A. Peavy, "Well Returns Analysis Causes Re-evaluation of HCl Theories" (paper SPE 14825, presented at the Society of Petroleum Engineers Formation Damage Control Symposium, Lafayette, LA, Feb. 26–27, 1986).

18. C. E. Shuchart, "HF Acidizing Returns Analysis Provide Understanding of HF Reactions" (paper SPE 30099, presented at the Society of Petroleum Engineers European Formation Damage Symposium, The Hague, Netherlands, May 15–16, 1995).

19. G. R. Coulter and A. R. Jennings, "A Contemporary Approach To Matrix Acidizing" (paper SPE 38594, presented at the Society of Petroleum Engineers Annual Technical Conference and Exhibition, San Antonio, TX, Oct. 5–8, 1997).

20. C. W. Crowe, "Precipitation of Hydrated Silica from Spent Hydrofluoric Acid: How Much of a Problem Is It?" *Journal of Petroleum Technology* (Nov. 1986): 1234.

21. L. J. Kalfayan and D. R. Watkins, "Discussion of Precipitation of Hydrated Silica From Spent Hydrofluoric Acid: How Much of a Problem Is It?" letter response, *Journal of Petroleum Technology* (Feb. 1987): 235.

22. R. D. Gdanski, "Kinetics of Tertiary Reactions of HF on Alumino-Silicates," *SPE Production & Facilities* (May 1998): 75–80.

23. R. D. Gdanski, "Fluosilicate Solubilities Impact HF Acid Compositions" (paper SPE 27404, presented at the Society of Petroleum Engineers International Symposium on Formation Damage Control, Lafayette, LA, Feb. 7–10, 1994).

24. R. D. Gdanski, "Fractional Pore Volume Acidizing Flow Experiments" (paper SPE 30100, presented at the Society of Petroleum Engineers European Formation Damage Symposium, The Hague, Netherlands, May 15–16, 1995).

25. R. D. Gdanski, "Kinetics of the Primary Reaction of HF on Alumino-Silicates," (paper SPE 37459, presented at the Society of Petroleum Engineers Production Operations Symposium, Oklahoma City, Oklahoma, March 9–11, 1997).

26. R. D. Gdanski, "Kinetics of the Secondary Reaction of HF on Alumino-Silicates" (paper SPE 37214, presented at the Society of Petroleum Engineers International Symposium on Oilfield Chemistry, Houston, TX, Feb. 18–21, 1997).

27. M. P. Walsh *et al.*, "A Description of Chemical Precipitation Mechanisms and Their Role in Formation Damage During Stimulation by Hydrofluoric Acid," *Journal of Petroleum Technology* (Sept. 1982): 2097–2112.

28. C. W. Crowe, "Evaluation of Agents for Preventing Precipitation of Ferric Hydroxide from Spent Treating Acid," *Journal of Petroleum Technology* (April 1985): 691–695.

part two

sandstone acidizing

Purposes of Sandstone Acidizing

4

Purposes of sandstone acidizing include:

- Perforation break down
- Near-wellbore formation damage removal

PERFORATION BREAK DOWN

It is sometimes necessary to break down perforations by temporarily pumping acid above fracturing pressure in order to initiate production or injection of a subsequent treatment, such as hydraulic fracturing. Typically, HCl acid is used in concentrations ranging from 5% to 20%, with 15% HCl being standard. As this is a routine procedure in certain areas, it will not be discussed in further detail.

NEAR-WELLBORE FORMATION DAMAGE REMOVAL

The primary purpose of matrix acidizing in sandstones is to remove formation damage caused by clay and other siliceous fine particles plugging near-wellbore permeability. Particles may be naturally occurring or may have been introduced into the formation during well operations. Damage can occur during drilling, completion, production, and stimulation operations.

Hydrofluoric acid (HF) is the only common acid that dissolves siliceous minerals appreciably. Therefore, sandstone acidizing formulations include HF or one of a variety of compounds that generates HF (HF precursor). The most commonly used formulations are mixtures of HCl and HF. These are referred to as "mud acid," from the early days of sandstone acidizing. Acid concentrations can vary from a low end (e.g., 3% HCl; 0.5% HF) to the high end (12–15% HCl; 3–5% HF, or more).

For sandstones with high carbonate mineral content (> 15–20% or more), HF should generally be avoided. Hydrochloric acid may be used alone in such cases. However, carbonates are often present in sandstones as grain cementation. Removal of carbonates with acid can diminish rock competence.

Hydrochloric acid is also applicable for removing certain scales, such as calcium carbonate, iron carbonate, iron oxides, and iron sulfide. Organic acids such as acetic and formic are sometimes used in place of HCl, especially in high-temperature applications, where HCl corrosion can be severe.

HF acid reaction in sandstones is fast, and only formation damage very near the wellbore can be treated effectively. This is a function of the very high surface area to volume ratios of siliceous minerals such as clays, feldspars, and zeolites, clays being highest. The large quartz grains have a very low surface area to volume ratio. HF reaction is controlled by these surface reaction kinetics. HF reacts preferentially with the high-surface-area particles. It usually spends within a short distance from the wellbore if these minerals are abundant, which they usually are. Treatment does not often exceed 1–2 ft beyond the wellbore (except in naturally fractured formations), and it can be much less.

Nevertheless, removal of very near-wellbore formation damage can result in several-fold increases in well productivity, as indicated by the substantial reduction in productivity resulting from severe, very near-wellbore damage, as shown in Table 4–1.

Depth of damage zone (inches)	Damaged Productivity / Undamaged Productivity — Damage zone permeability ÷ undamaged permeability				
	0.5	**0.2**	**0.1**	**0.05**	**0.02**
0	1.0	1.0	1.0	1.0	1.0
1	0.95	0.90	0.80	0.65	0.40
2	0.93	0.82	0.65	0.49	0.25
4	0.91	0.74	0.54	0.37	<0.1
6	0.89	0.68	0.47	0.3	~0
12	0.85	0.60	0.40	0.22	~0

Table 4–1. *Effect of Formation Damage on Production (Radial Flow)*

If damage extends beyond a few inches, and is severe, the productivity increases possible are dramatic. For example, removal of damage impairment of 95% to 6" beyond the wellbore will result in a greater than 3-fold increase in production. If permeability impairment was 98%, damage removal might realize a 10-fold productivity increase, or more.

On the other hand, if skin damage is not present, matrix acidizing cannot be expected to result in significant stimulation, as suggested by the information shown in Table 4–2.

To double production, acid would have to effectively contact formation beyond 10 ft, which is not realistic in a sandstone. If stimulation treatment is desired in an undamaged sandstone, fracturing would be called for.

Sandstone acidizing can be a very successful well stimulation method. However, the risk of failure is fairly high. Fortunately, there are a limited number of reasons why sandstone acidizing treatments fail. Success begins with the selection of a viable acidizing candidate well. Many poorly producing wells are not viable acidizing candidates. Once a viable candidate

Zone of Effective Acid Contact (feet)	Productivity After Acidizing / Undamaged Productivity
1	1.2
2	1.4
4	1.6
6	1.7
10	1.9
20	2.6
40	2.9
60	3.3
100	4.1

Table 4–2. *Effect of Matrix Acidizing on Radial Flow Potential in an Undamaged Sandstone (40-Acre Spacing Example)*

is selected, a systematic approach to the selection of fluids, additives, and acid placement technique must be taken. On-site quality control and posttreatment evaluation help to ensure successful results and improved future treatments.

Why acid jobs fail is discussed in chapter 5. The subject of chapter 6 is preventing failure and increasing the success rate through a systematic approach to sandstone acidizing treatment design.

5 Why Sandstone Acid Jobs Fail

The popular perception is that acid jobs have a high failure rate. That is unfair, as success is very site-specific. In any case, acidizing is often considered a "hit-or-miss" prospect. This reputation is well deserved, if based on results alone, and if the factors leading up to, and during, the acid treatment are not considered.

In reality, there are a limited number of reasons, or controllable causes, for sandstone acidizing treatment failure, and all can be avoided with proper treatment planning and execution. This is not to say that a success rate of 100% is readily attainable. It may be possible in a particular field, or with a limited selection of wells, but no stimulation method is foolproof. Acidizing is no exception.

There are reasons why stimulation treatments fail that cannot be controlled. These include unforeseen mechanical problems, especially downhole with the well and with tools, as well as with on-site stimulation equipment. This can be more frequent offshore with more compact, higher technology equipment designed for remote application.

Another reason for failure that is part of the risk with stimulation, including acidizing, is the possibility that a poor stimulation candidate well is treated. This is usually because the information needed to fully evaluate the well simply does not exist. However, some kind of stimulation treatment may be attempted; either as a last resort or because it is required following drilling and completion, for example. It should be understood that sometimes a well just has no chance of being a good well, stimulated or not.

Fortunately, there are many good wells, and good acid stimulation candidates. In such cases, most potential causes of acid treatment failure can be addressed through a systematic approach to treatment candidate selection and treatment design.

First, we must identify and understand the causes of sandstone acidizing treatment failure that are avoidable. These should always influence the steps we take in the treatment decision and design process.

Industry experts such as Harry McLeod and George King have done us great service by identifying and listing the common causes of sandstone acidizing treatment failures in their teachings. As it turns out, most acidizing treatment failures can be explained by one or more of the following:

1. Treating a well that has high skin, but no damage

2. Using acid on a formation that was not adequately perforated

3. Using the wrong type of acid to remove the damage

4. Using improper acid volumes and/or acid concentrations for the formation mineralogy

5. Using dirty water to mix preflush or overflush stages

6. Failure to clean acid or water tanks

7. Additive overuse or misuse

8. Pumping the acid job above fracturing pressure (with exceptions)

9. Shutting in the acid treatment too long before producing back

Understanding that most treatment failures are due to one or more of these reasons simplifies the process. It also can ease the mind when sandstone acidizing treatment design considerations seem hopelessly complicated.

TREATING HIGH-SKIN WELLS WITH NO DAMAGE SKIN

If formation damage (skin damage) does not exist in a matrix formation, then acidizing cannot be expected to impart significant stimulation. This is a mathematical limitation in radial flow through a matrix. If acid-removable skin = 0, or is insignificant, skin reduction to a negative value will not increase flow through the matrix enough to justify treatment.

This is more the case with sandstone matrix formations. In carbonates, and perhaps in naturally fractured reservoirs, there is a greater stimulation potential through an undamaged formation. However, just the same, significant stimulation cannot be expected in any undamaged formation case: sandstone or carbonate, matrix or naturally fractured.

FORMATION NOT ADEQUATELY PERFORATED

In a new well (or in a recompleted or reperforated well), if perforations are incomplete, a positive skin value will be present. However, it is present as a "pseudoskin," not as skin due to acid-removable formation damage. The total skin effect may be written as:

$$s_{total} = s_{partial\ completion} + s_{perforation} + s_{damage} + pseudoskins$$

The last term on the right refers to the large number of pseudoskin factors that may exist, including phase-dependent and rate dependent effects.[1] The term $s_{partial\ completion}$ gives the skin value due to partial completion and slant.[2] The $s_{perforation}$ represents the skin effect due to incomplete perforations.[3] The term s_{damage} is the only skin that acidizing can address.

Acidizing cannot address insufficient, inadequate, incomplete, or inefficient perforating. Reperforating is the only likely viable option in this case. Hydraulic fracturing may be used if fluid entry is possible.

USE OF INCORRECT ACID

Acid-removable damage may be present (or determined to be present), but the right type of acid must be used to remove the damage. For example, hydrochloric acid will not dissolve plugging solids such as clays and other siliceous fines. HF acid should not be used to remove calcium carbonate solids.

USE OF INCORRECT ACID VOLUMES OR CONCENTRATIONS

Certain formations are very sensitive to acid volumes and concentrations, especially with respect to HF acid. Sands with high clay content, for example, may be damaged by use of high-strength HF, as reprecipitation of reaction products near the wellbore dominate. High-strength HCl solutions, or even acid mixtures containing HCl, can be very damaging to sandstones containing high levels of iron chlorite clay and certain zeolites (aluminosilicate minerals). Analcime is most acid-sensitive. Others may or may not be acid-sensitive at all. Acid-mineralogy guidelines are presented and discussed in chapter 6.

USE OF DIRTY WATER TO MIX PREFLUSH OR OVERFLUSH STAGES

This requires no explanation. All fluids must be as clean as possible. Source waters used to mix stimulation fluids must be adequately filtered. See chapter 16 for more information about on-site acidizing quality control practices.

FAILURE TO CLEAN ACID OR WATER TANKS

Again, quality control practices will address this issue. Fluids, as well as mixing and holding tanks, must be as clean and solids-free as possible. It is still surprising to see how often tanks are not adequately cleaned prior to acidizing. This is especially a problem during "good times," when well workover and stimulation services are needed on a frequent basis, and competition is keen. Tanks and equipment are hustled from one location to another, with thorough flushing and cleaning sometimes skipped in the interest of starting a new treatment, or completing one and moving on to the next.

A perfectly reasonable and potentially effective acid treatment design can be ruined by pumping dirty fluids or fluids from dirty or rusty tanks. These can carry debris or high levels of dissolved iron that will deposit or reprecipitate in the formation, plugging perforations or pore spaces near the wellbore.

ADDITIVE OVERUSE OR MISUSE

Additives are discussed in chapter 6. Excessive use of additives or mixing additives that are not compatible with other additives in acid can absolutely ruin a treatment. There are many possibilities and additive combinations that can be categorized as additive overuse or misuse. Among the most common are excessive concentration of potentially oil-wetting surfactants, such as corrosion inhibitors or clay stabilizers. Also, mixing certain additives, such as iron-control agents, beyond their solubility limits can happen. If so, insoluble solids are unnecessarily pumped into the formation.

Different iron-control additive types have varying solubility limits in different acids, and the limits vary with acid concentration, sometimes increasing, sometimes decreasing. Additive solubility limits, as well as compatibilities, must be carefully checked by the service company laboratory before finalizing an additive program.

PUMPING THE ACID JOB ABOVE FRACTURING PRESSURE

Sometimes acid must be pumped above fracturing pressure, just to break down perforations and initiate flow. However, it is generally accepted that sandstone acidizing must take place in the matrix—within the pore spaces—to impart stimulation.

There are occasions where treating above fracturing pressure may make sense; for example, in certain naturally fractured formations. There may also be certain sandstones (in rare cases) in which acid-etching is possible, similar to that achievable in carbonate formations. This is not necessarily a desirable effect in a sandstone. The formation would have to be competent enough to sustain or support a conductive channel. It is not a likely phenomenon in a sandstone, and cannot be predicted with confidence. It does bear consideration, at least in the back of the mind, especially with particularly hard rock.

SHUTTING IN THE ACID TREATMENT TOO LONG

Shutting in an acid treatment, especially HF acid in a sandstone near the wellbore, increases the chance of damage in the formation from precipitation of HF reaction products. Acid, especially spent HF, should be produced back out immediately, or as soon as possible. If immediate turnaround cannot be accomplished, then acid should be kept moving, and a healthy overflush should be employed. This is so that reprecipitation of acid reaction products, which inevitably takes place, will be far enough beyond the near-wellbore that its effect on radial permeability is insignificant.

In carbonate formations, there is usually not much concern with reprecipitated species. However, if iron is present or H_2S gas is expected, certain scales and solids reprecipitation can occur. These can be substantially damaging if fluids are shut in and allowed to remain static for long periods of time (a few hours or more).

Some shut-in period is often unavoidable anyway, as a result of logistics and the time it takes to repipe lines and turn a well around. Sometimes a well does not produce back immediately, and assistance is required, such as swabbing and jetting with nitrogen. Hopefully, the shut-in period can be limited to 2–8 hours. Beyond that range there is cause for concern, especially with lower permeability sandstone formations with complex mineralogy, conducive to spent acid reprecipitation reactions.

Before we delve into the systematic approach to acid treatment, well candidate selection, and acid treatment design, it is important to have a feel for the common reasons for acid treatment failure. It is also helpful to understand the causes of treatment failure that are usually avoidable. Acid treatment design cannot be cookbooked. One should not attempt to reproduce a boilerplate acid job procedure throughout an entire field, as desirable as this may seem.

However, the approach can be reproduced, and should be. By following a systematic approach to acid treatment design, the controllable potential causes of failure can be addressed and eliminated to a great extent. The chance of success is thereby greatly increased.

REFERENCES

1. M. J. Economides and K. G. Nolte, editors, *Reservoir Stimulation* 2d ed., (Schlumberger Educational Services, 1989).

2. H. Cinco-Ley, H. J. Ramey, Jr., and F. G. Miller, "Pseudoskin Factors for a Partially Penetrating Line-Source Well" (paper SPE 5589, Society of Petroleum Engineers, 1975).

3. M. H. Harris, "The Effect of Perforating on Well Productivity," *Transactions*, American Institute of Mechanical Engineers (1966) 237 (Sec. I): 518–28.

6 Six Steps To Successful Sandstone Acidizing

As discussed in the previous chapter, it is important to understand that there are a limited number of reasons why sandstone acidizing treatments do not succeed. This understanding is the first step in overcoming the perception of risk-taking and frequent failure in sandstone acidizing. This understanding is also needed to take an interest in the exciting prospects of acidizing. In order to make the most of acidizing, acid treatment design must be approached as a process, rather than as an afterthought or as a last resort, as it often is.

The general approach to be taken with sandstone acidizing, and all acidizing treatments for that matter, is as follows:

1. Select an appropriate stimulation candidate well
2. Design an effective treatment
3. Monitor the treatment for subsequent improvement

In selecting an acidizing candidate well, it is first important to analyze well performance and the reservoir properties. In analyzing well performance and assessing a well as an acidizing candidate, the following questions should be asked:

1. Is the well performing poorly?
2. If so, is poor reservoir quality the cause?
3. Or, is formation damage the cause?
4. If so, what are the types of damage (damage contributors)?
5. What is the severity of damage (skin)?
6. What is the location and depth of penetration of the damage?
7. Can acid remove the damage?
8. If so, what type of acid will remove the damage?
9. Will the formation be compatible with this acid?
10. If not, what alternative nonacid treatment might remove the damage?

Well performance is largely dependent on reservoir quality and the nature of the formation matrix, as well as on the presence and type of damage.

The sandstone acidizing design approach to be taken is introduced and presented in this chapter as a six-step process, as inspired by the work and teachings of Harry McLeod.[1] At the end of the chapter, a list of references for recommended reading on the subjects related to sandstone acid treatment design is given.

The six-step process to successful sandstone acidizing is as follows:

1. Determine the presence of acid-removable skin damage
2. Determine appropriate fluids, acid types, concentrations, and treatment volumes
3. Determine proper treatment additive program
4. Determine treatment placement method
5. Ensure proper treatment execution and quality control
6. Evaluate the treatment

This process first involves investigation and assessment of the stimulation candidate well to determine if sufficient productivity improvement is, in fact, possible. Then it is necessary to evaluate the damage present, as well as the reservoir quality and mineralogy, to determine the appropriate fluids needed. These considerations include acids, acid types, concentrations, and volumes. The proper additive program must then be determined, avoiding excessive or unnecessary additives.

It is then necessary to determine if a treatment placement method must be employed: mechanical, chemical diversion, or a combination. The final steps are taken after the design procedure is determined and the acid job is to be pumped. Proper treatment execution and quality control practices must be in place to ensure success and safe practices in the current treatment and in subsequent treatments. Finally, measures to evaluate the treatment completed should be implemented to impact future design.

Sometimes it seems that sandstone acidizing treatment design is overwhelming. It may seem that there are too many variables, too many issues to worry about, and too many choices. It is true that there are many variations to the acids, their concentrations, and volumes, the additive choices, and the number of steps in an acidizing procedure. However, bear in mind that all sandstone acid treatments are variations of the following maximum step procedure:

1. Crude oil displacement (solvent) stage
2. Formation water displacement stage
3. Acetic acid stage
4. HCl preflush stage
5. Main acid (HF) stage
6. Overflush stage
7. Diverter stage
8. Repeat steps 2–7 (as necessary)
9. Repeat steps 2–6
10. Final displacement stage

Most treatments will not require this number of steps. A gas well acid treatment, for example, will probably not need step 1 (crude oil displacement stage). The exception is if condensate is produced, and it is desirable to separate it from contact with stimulation fluids. Step 3 (acetic acid stage) may only be needed in a formation rich in iron minerals; or it may be the preflush step itself, without the need for step 4 (HCl preflush stage).

Other steps may or may not be needed, depending on the type of damage present, the nature of the formation fluids, and the formation itself. Each step can be varied, as well, with respect to fluid types, acid types, concentrations, additives, diverters, and so on.

In the systematic approach to sandstone acid treatment design, it is best to work within this general design framework that fits all cases. It is the variation within this framework that is subject to our ideas, choices, and creativity.

STEP ONE: DETERMINE PRESENCE OF ACID-REMOVABLE SKIN DAMAGE

A well producing from a sandstone formation is only a candidate for acidizing if acid-removable skin damage is present. With respect to acid treatment design, it is not only important to determine whether skin damage (positive skin) is present, but if so, if the damage can be removed with an acid treatment.

If a well is not damaged, or if it is, but the damage is not acid-removable, there should be no expectations from an acidizing treatment. Such a well should not be acidized. This is because acidizing does not decrease skin very much below zero, as can be shown from the radial flow mathematics.[2, 3] However, an undamaged well may respond to hydraulic fracturing, if feasible from both a mechanical and reservoir standpoint.

The point is that acidizing in sandstones only addresses skin, which can have a drastic effect on well productivity. As explained in chapter 3, well production rate, q, can be defined by Darcy's law for steady-state liquid flow in a radial reservoir as a function of permeability, k, and skin, s, among other factors:

$$q = 7.08kh\,(p_e - p_{wf}) \,/\, B\mu\,[(\ln r_e/r_w) + s]$$

Again, skin, s, is a multicomponent total skin (s_t) term, composed of different skin "contributors," as follows:

$$s_t = s_{c+\phi} + s_p + s_d + \Sigma_{pskins}$$

where

$s_{c+\phi}$	is the skin due to partial completion
s_p	is the skin due to incomplete perforations
s_d	is the skin due to damage
Σ_{pskins}	is the variety of pseudoskin factors (phase- and rate-dependent effects)

It is the aim of sandstone acidizing treatments to reduce that portion of the total skin (s_t) due to damage (s_d). Damage skin must be present, but it must be acid-removable, as manifested in the wellbore, the perforations, and/or within the formation.

In evaluating a well producing from a sandstone reservoir as a stimulation candidate, skin must be measured, or at least assessed as best as possible, to select the proper treatment course (or nontreatment).

ACID-REMOVABLE SKIN DAMAGE

Generally speaking, acid-removable skin damage is a reduction in permeability caused by plugging or constriction in pore throats, which can be removed by acid. Skin that is not acid-removable also includes changes to the pore structure resulting in increased resistance to flow, such as wettability effects. Resistance to flow also may result from plugging by acid-insoluble materials, such as certain scales, paraffin (wax), and asphaltenes.

Acid-removable formation damage can occur during any well operation. Table 6–1 summarizes types of acid-removable damage. HF acid mixtures can be used to remove the types of damage listed, unless otherwise noted.

Well Operation	Damage Mechanism
Drilling	Mud solids invasion Mud filtrate invasion
Cementing	Filtrate invasion (high pH effect)
Perforating	Compaction of perforated zone Formation debris
Production	Inorganic scale plugging [a] Calcium carbonate Iron scales (varying acid solubilities) Fines migration
Workover	Solids invasion Clay swelling; migration (incompatible brine)
Stimulation	Release and migration of fines [b] Precipitation of solids formed from the reactions of stimulation fluids with formation minerals or fluid [c] Polymer damage (frac fluid) [d] Formation wettability alteration [e]

(a) Removed by HCl acid alone; acetic, formic acids alone or in combination with HCl, or EDTA chelating complexes are alternatives to HCl

(b) Aluminosilicate minerals (clays, feldspars, zeolites); silica (quartz fines)

(c) May be insoluble precipitates

(d) May preclude use of HF; HCl only usually sufficient

(e) Caused by additives; treatment with surfactant required

Table 6-1. *Acid-Removable Skin Damage*

STEP TWO: DETERMINE APPROPRIATE FLUIDS, ACID TYPES, CONCENTRATIONS, AND TREATMENT VOLUMES

Once it has been determined that acid-removable formation damage is present, and that treatment is mechanically feasible, the proper acid type, acid volume, and acid concentrations must be determined. As mentioned earlier, the maximum-step conventional HF-acid treatment design, with typical volumes per foot of perforations or zone, is given in Table 6–2. The minimum-step conventional HF-acid treatment design is given in Table 6–3.

Stage	Volume
1. Crude oil displacement (solvent) stage	10–75 gal/ft
2. Formation water displacement stage	15–25 gal/ft
3. Acetic acid stage	25–100 gal/ft
4. HCl preflush stage	25–200 gal/ft
5. Main acid (HF) stage	25–200 gal/ft
6. Overflush stage	
7. Diverter stage	10–250 gal/ft
8. Repeat steps 2–7 (as necessary)	
9. Repeat steps 2–6	tubing volume +
10. Final displacement stage	

Table 6-2. *Maximum-Step Conventional HF Acid Treatment Design*

Acid Preflush	25–200 gal/ft
Main Acid (HF) Stage	25–200 gal/ft
Overflush	10–250 gal/ft

Table 6-3. *Minimum-Step Conventional HF Acid Treatment Design*

Tubing pickling stage

If possible, the injection string (production tubing, drill pipe, and coiled tubing) should be "pickled" prior to pumping the acid treatment. Inhibited 5% HCl or special pickling solutions may be used. Service companies have pickling solutions containing an iron-control agent, and sometimes an aromatic solvent with surfactant. A basic pickling solution is 5% HCl containing an iron-control agent and corrosion inhibitor. Sometimes, because of location space or tank volume limitations, the same acid mixed for use in the HCl preflush may be used as the pickling solution.

Crude oil displacement stage (optional)

This stage is applicable in oil wells in which the crude may not be compatible with acid mixtures used. It is also applicable where additives to ensure compatibility cannot be determined with confidence. Sometimes this stage may also be beneficial in gas/condensate producers. However, one should generally avoid pumping hydrocarbon solvent into a gas-bearing formation.

If condensate is appreciable, or if it is known to accumulate near the wellbore, then a solvent stage is acceptable. An aromatic solvent, such as xylene, is used. Typically, 10–75 gal/ft are adequate, but a preferable volume is 25–50 gal/ft. Xylene or another aromatic solvent may be combined with acid and surfactants. It is preferable to keep them separate, if possible. The purpose is to displace oil away from the acid stages to prevent sludge or emulsion, or to remove paraffin/asphaltene wellbore deposits.

Formation water displacement stage (optional)

A water displacement stage ahead of the standard acid preflush is also optional. The recommended fluid is 2–7% ammonium chloride (NH_4Cl) solution, with 5% NH_4Cl preferable for most applications. Recommended volume is 15–75 gal/ft. Typical volume is 25–50 gal/ft. The purpose of this stage is to displace formation water containing appreciable bicarbonate and sulfate ions. A concentration greater than 1000 parts per million (ppm) of these ions is considered appreciable.

By including a nonacid water displacement stage, the precipitation of calcium carbonate and calcium sulfate scales can be avoided. These scales form when spent HCl (calcium chloride in solution) mixes with formation brine.

Acetic acid stage (optional)

A sandstone to be treated can be high in iron compounds such as siderite (iron carbonate), pyrite (iron sulfide), iron oxides, and iron chlorite clay, etc. If this is the case, a separate acetic acid stage may be needed to reduce iron precipitation potential. Recommended acetic acid volume is 25–100 gal/ft. Acetic acid reacts with carbonates and reduces reprecipitation of dissolved iron compounds through complexation.

If the formation is high in iron content and high in carbonate content, both an acetic acid stage and a hydrochloric acid preflush stage may be needed. If iron compounds are present but the carbonate content is not so high (< 5%), the acetic acid stage can serve as the acid preflush, precluding the need for step 4 (HCl preflush). If iron mineral content is low, this separate acid preflush stage is not necessary. The recommended stage is 10% acetic containing 5% ammonium chloride (NH_4Cl) for acetic acid stages. Ammonium chloride is added for clay stability. Acetic acid without the salt added may slightly swell clays.

ACID PREFLUSH

The main purpose of the preflush is to dissolve carbonate minerals in the formation prior to injection of the main HF acid mixture. HF acid reacts with carbonates, such as calcium carbonate and magnesium carbonate, to form insoluble calcium and magnesium fluorides. If a separate water displacement stage is not employed, the acid preflush serves the additional purpose of displacing formation water from the main HF acid stage. If given the opportunity, spent HF acid will further react with sodium, potassium, and calcium ions in formation brine to form insoluble precipitates that can cause severe plugging in the formation.

The standard preflush is hydrochloric (HCl) acid, usually 5–15%. Organic acids, such as acetic and formic, can also be used by themselves, in combination with each other, or in combination with HCl. Organic acids are especially useful in high-temperature applications, because they are less corrosive than HCl. Formic/acetic blends are popular for this reason. Acetic acid is also useful by itself, or in combination with HCl in formations with a high-iron content (see acetic acid stage, above).

If the preflush cannot be injected because of very severe damage, it may be necessary to break down the formation by pumping above fracturing pressure. This is acceptable as long as injection is returned to matrix rate (below fracturing pressure) as soon as possible. An alternative is to forego the preflush and break down the severely damaged formation with the main HF acid phase initially.

If break down is successful, a return to the preflush injection can be made. Alternatively, the treatment can continue with the HF stage if it appears damage is being removed successfully, signified by an injection pressure decrease. It is possible that if formation damage is so severe that preflush acid cannot enter the formation, the additional damage caused by injecting HF acid without the preflush will not be significant.

In any case, every effort should be made to inject the preflush and main acid phases below fracturing pressure.

MAIN ACID

The purpose of the main acid stage is to dissolve siliceous particles that are restricting near-wellbore permeability, plugging perforations or gravel packs. The main acid phase is a mixture of hydrochloric (HCl) and hydrofluoric (HF) acids. Common HCl-HF mixtures include:

- 12% HCl, 3% HF (regular strength mud acid)
- 13.5% HCl, 1.5% HF
- 7.5% HCl, 1.5% HF
- 6% HCl, 1.5% HF
- 9% HCl, 1% HF
- 6.5% HCl, 1% HF
- 3% HCl, 0.5% HF

Volumes may range from 25–200 gal/ft or more. Volume is somewhat arbitrary, but should have a logical dependence on formation permeability, acid sensitivity, and type and severity of damage. Retarded acid mixtures and other HF-generating systems are sometimes used in place of, or in addition to, conventional HF acid stages. For the most part, a properly designed conventional treatment with HCl-HF acid mixtures, or with organic acid/HF mixtures, will stimulate damaged sandstone formations.

Risks associated with acidizing, such as fines migration, precipitation of reaction products, and rock deconsolidation normally can be minimized with proper volumes and concentrations of acids used. Therefore, in most cases, what can be achieved with a "novel" acidizing system can be achieved with an appropriate conventional HF treatment.

Novel acidizing systems, which come and go, serve to increase service company profits more than they do to enhance acidizing response. However, retarded HF systems can be uniquely beneficial in exceptional cases or niche applications. Such cases might include the need for deep damage removal and treatment of highly acid-sensitive sands (or sands with certain unusual mineralogies). Certain service company retarded systems are less corrosive and easier to mix on site, as well, and may be considered with those features in mind.

With conventional HF acid mixtures, HF concentration should generally not exceed 3%, except under special circumstances. Such circumstances might include geothermal features, certain "clean" sands, certain conglomeritic sandstones, and other consolidated sands.

HF acid shut-in time should be minimized to reduce reprecipitation of reaction products. One should be skeptical of HF treatments that require an intentional shut-in period as part of the procedure, even in high permeability formations. If HF shut in is to exceed one day, the overflush should be designed to displace spent HF 3–5 ft from the wellbore.

Do not use NaCl, KCl, or $CaCl_2$ brines in any HF acid treatment stages, or in any stage immediately preceding or following HF stages.

The HF acid stage should be pumped below fracturing pressure. Acid fracturing is really not applicable in sandstones, as it is in carbonates. Like all stimulation scenarios, there are exceptions. However, for the most part, etching of the fracture faces for development of flow conductivity normally does not take place in sandstones.

Acid fracturing with HF may have limited application in naturally fractured formations or in very soft sands that cannot be "propped." When severe formation damage is encountered or when perforations need to be broken down, the main acid stage injection may briefly be allowed to exceed fracturing pressure. However, once injection is established, matrix treatment should be resumed.

Overflush

The purpose of the overflush is to displace the HF acid phase away from the wellbore. By doing so, precipitation reactions that inevitably take place will only occur well away from the near-wellbore region, where the effect on productivity will be insignificant.

The overflush should normally be 2–8% ammonium chloride solution. An overflush of 5% NH_4Cl is recommended. Weak HCl or acetic acid (3–5%), filtered diesel, or even lease crude, can be used as overflush fluids. If acetic acid alone is used, 5% NH_4Cl should be added to improve clay stability. The use of ammonium chloride/acetic acid mixtures has gained popularity, especially in the Gulf of Mexico. In gas wells and sometimes in extremely water-sensitive formations, nitrogen is an effective overflush. If oil-soluble particulates have been used to divert acid, an aromatic solvent stage preceding the ammonium chloride overflush may be needed.

Overflushes containing acid maintain a low pH near the wellbore, thereby preventing certain precipitation reactions, such as iron precipitation, which are dependent on fluid pH. Diesel or lease crudes are used if reestablishing oil saturation near the wellbore is required for inducing normal productivity.

Brines containing sodium, potassium, or calcium ions should never be used to overflush HF acid. Sodium and potassium ions react with spent HF to form insoluble precipitates (sodium and potassium fluosilicate salts). Calcium ions react to form insoluble calcium fluoride.

ACID TREATMENT VOLUME AND RATE GUIDELINES

Broad volume ranges were shown previously in the general treatment procedures. Recommended acid treatment fluid volumes are not as broad and are summarized in Table 6–4.

	Permeability Range (mD)			
	1–10	**10–25**	**25–100**	**100 +**
Preflush	25	25–50	35–75	50–150
HCl-HF	25–50	25–50	75–100	75–150
Overflush	Oil wells: > HCl-HF volume			
Gas wells:	< HCl-HF volume; or use N_2 gas only			

Table 6-4. *Recommended Fluid Volumes for Basic Treatment (gal/ft of perforations)*

Lower permeability sands should be treated with lower acid volumes because of the increased sensitivity to damage caused by the treatment. Also, if a formation contains high clay contents, especially the swelling variety (smectite, illite-smectite), volumes may need to be reduced to a further extent.

ACID INJECTION RATES

The rates of acid injection are dictated by allowable injection pressure. As presented by Piot and Perthius, the maximum injection rate that will not fracture the formation may be estimated from Darcy's radial flow equation, represented as:[2]

$$q_{imax} = 4.917 \times 10^{6} kh\,[(g_f \times H) - \Delta p_{safe} - p]\,/\,\mu B(\ln r_e/r_w + s)$$

where

q_{imax}	is the injection rate (bpm)
k	is the permeability of the undamaged formation (mD)
h	is the net thickness of the formation (ft)
g_f	is the fracture gradient (psi/ft)
H	is the depth (ft)
Δp_{safe}	is the safety pressure margin (200 to 500 psi)
p	is the reservoir pressure (psi)
μ	is the_viscosity (centipoise)
B	is the formation volume factor (very near unity)
r_e	is the drainage radius (ft)
r_w	is the wellbore radius (ft)
s	is the skin factor (dimensionless)

This equation is a simplified radial flow equation, in that it only accounts for an incompressible injected fluid in a homogenous formation. It does not take multiphase flow effects into account. This representation of Darcy's equation should only be used as a guideline for setting the initial treatment injection rate. It is good practice to limit acid treatments in sandstones to several hours, at the most, whenever possible. Long HF residence time may be expected to increase precipitation of acid reaction products. HF contact time should be limited to 2–4 hours per stage, if possible.

The HCl (or organic acid) preflush should be at least 50–100% of the HCl-HF volume. If solubility of the formation in HCl is less than 5%, then the preflush should be 50% of the HCl-HF volume. If the solubility in HCl is between 5–10%, then the preflush should be 100% of the HCl-HF volume. In formations with HCl solubility greater than 10%, the preflush should be 150% of the HCl-HF volume.

Low acid volumes should be used in poorly consolidated formations, low permeability formations, and in formations with high clay content. In poorly consolidated sands, excessive HF can cause sloughing or collapse of perforations, and possibly sand production. Excessive HF in low-permeability formations and in formations with high clay content can cause severe plugging due to reprecipitation of dissolved solids. Formations with native permeability less than 10 mD can be acidized with care. However, hydraulic (propped) fracturing may often be the stimulation method of choice.

It is important to note that in sandstone acidizing, more acid is not necessarily better. In fact, in most cases, the opposite is true. The formation penetration distance of live acid is usually much less than 1 ft. However, near-wellbore formation damage is substantially removed during the initial contact of the formation with acid. Injection of higher volumes of acid may only serve to dissolve and weaken the matrix, rather than remove residual damage. Therefore, acid volumes must be moderated.

There are few cases requiring greater than 150–200 gal/ft of HF acid. These are limited to fractured formations, such as shales, where high volumes of acid can open fracture networks deeper in the formation.

In treatments where on-site, real-time skin measurement is monitored, acid injection should be cut off once it appears that skin is removed. This is not a common practice, as it is generally undesirable on the part of the service companies. This is because unused acid must be taken away and disposed of. However, it is acceptable if the possibility is understood beforehand, and the treatment can be fairly priced accordingly. Continuing to pump HF after damage is removed is counterproductive and may begin to damage the formation again, undoing the good that was accomplished.

In any case, if formation damage is so severe that volumes of 150–200 gal/ft of HF acid cannot penetrate the damage, then hydraulic (propped) fracturing must be considered.

ACID TREATMENT CONCENTRATION GUIDELINES

Selection of acid concentrations must be based on the formation characteristics. Knowledge of permeability, porosity, and mineralogy is imperative. Amounts and types of clays and feldspars are especially important to ascertain. This information can be obtained through X-ray diffraction analysis.

However, the location of minerals is of greatest importance. SEM (scanning electron microscopy) and thin-section analysis are additionally useful in identifying locations of quartz, clays, feldspars, carbonates, and other minerals. These are all factors in acid treatment design. Expertise available in the major service companies should be contacted to help in treatment development. Input from company geologists should also be sought.

In 1984, McLeod introduced his acid use guidelines to the industry.[1] This was a breakthrough in formation-based acid treatment design. The guidelines focused on mineralogy, which was, and still is, often overlooked. The original McLeod guidelines are shown in Table 6–5.

Since their original introduction, McLeod's guidelines have been modified by service companies to fill certain so-called gaps. These gaps included permeability ranges (10–100 mD) and higher temperature conditions.[4] Also, the original guidelines have been augmented with respect to certain mineral sensitivities. Specifically, these modifications included consideration for the presence of acid-sensitive zeolites and the use of higher HCl:HF ratios to reduce reprecipitation reactions. Particular emphasis was placed on including 9% HCl-1% HF for higher feldspar contents.

An example set of nonproprietary guidelines, based on previous modifications, as well as my own, are given in Table 6–6.

Formation	Main acid	Preflush
Solubility in HCl > 20%	Use HCl only	
High Permeability (>100 mD)		
High quartz (>80%); low clay (<5%)	12% HCl-3% HF	15% HCl
High feldspar (>20%)	13.5% HCl-1.5% HF	15% HCl
High clay (>10%)	6.5% HCl-1% HF	sequestered 5% HCl
High iron chlorite clay	3% HCl-0.5% HF	sequestered 5% HCl
Low Permeability (10 mD or less)		
Low clay (<5%)	6% HCl-1.5% HF	7.5% HCl or 10% acetic
High chlorite	3% HCl-0.5% HF	5% acetic

Table 6-5. *Original McLeod Sandstone Acidizing Use Guidelines*

For HCl-HF mixtures shown above, a small amount of acetic acid (e.g., 3%) may be added to reduce precipitation of aluminum fluoride compounds.[5] This is discussed more later when additives are discussed.

Unfortunately, since McLeod introduced his original guidelines, further modifications, particularly those made by the service companies, have been largely self-serving. For the most part, these have been used in order to promote current product preferences and proprietary acid systems. Most importantly, such guidelines, contrived or not, are too often misused.

It is tempting to generalize sandstone acidizing treatments based on mineralogy alone and take the guidelines too far. It is impossible to make such guidelines all-inclusive. The location of minerals in the matrix and the type and severity of formation damage are equally important or more important than the bulk mineralogy itself. Sandstone acidizing treatment design should be dictated by a combination of these considerations.

We must be open to exceptions and be willing to go against conventional wisdom. Set guidelines diminish our capacity to explore options and to take reasonable risks. They also prevent us from seeking advice from available experts.

Formation	Main acid	Preflush
Solubility in HCl > 15–20%	avoid use of HF, if possible	
Calcite or dolomite	15% HCl only[1]	5% NH_4Cl
High iron carbonate (ankerite, siderite)	15% HCl + iron control[1,2]	5% NH_4Cl + 3% acetic
High Permeability (>100 mD) [3,4]		
High quartz (>80%); low clay (<5%)	12% HCl-3% HF	15% HCl
Mod. clay (<5–8%); low feldspar (<10%)	7.5% HCl-1.5% HF	10% HCl
High clay (>10%)	6.5% HCl-1% HF	5–10% HCl
High feldspar (>15%)	13.5% HCl-1.5% HF	15% HCl
High feldspar (>15%) and clay (>10%)	9% HCl-1% HF	10% HCl
High iron chlorite clay (> ~8%)	3% HCl-0.5% HF or, 10% acetic-0.5% HF	5% HCl 10% acetic + 5% NH_4Cl
Medium Permeability (10–100 mD) [3,4]		
Higher clay (>5–7%)	6% HCl-1.5% HF	10% HCl
Lower clay (<5–7%)	9% HCl-1% HF	10% HCl
High feldspar (>10–15%)	12% HCl-1.5% HF	10–15% HCl
High feldspar (10–15%) and clay (>10%)	9% HCl-1% HF	10% HCl
High iron chlorite clay (> ~8%)	3% HCl-0.5% HF 10% acetic-0.5% HF	5% HCl 10% acetic + 5% NH_4Cl
High iron carbonate content (>5–7%)	9% HCl-1% HF 5% HCl-0.5% HF (k<25 mD)	10% HCl 10% HCl
Low Permeability (1–10 mD) [3,4,5]	Consider hydraulic fracturing first	
Low clay (<5%); low HCl sol. (<10%)	6% HCl-1.5% HF	5% HCl
High clay (>8–10%)	3% HCl-0.5% HF	5% HCl
High iron chlorite clay (>5%)	10% acetic-0.5% HF	10% acetic + 5% NH_4Cl
High feldspar (>10%)	9% HCl-1% HF	10% HCl

Very low permeability (<1mD) Avoid HF acidizing; non-HF matrix stimulation (dictated by damage) or hydraulic fracturing is preferred

(1) *Location of carbonate in matrix is important; it may be possible to include HF in naturally fractured formations with high carbonate content.*

(2) *HCl can be replaced by acetic or formic acid—partially or completely—especially at higher temperatures (250–300 °F).*

(3) *If zeolites (analcime) are present (> ~3%), consider replacing HCl with 10% citric acid or special service company organic acids.*

(4) *For higher temperatures (> 225–250 °F), consider replacing HCl with acetic or formic acid.*

(5) *Although fracturing may be preferable, low permeability, low clay-containing sands may respond favorably to HF acidizing—contrary to conventional wisdom.*

Table 6-6. *Conventional Sandstone Acidizing Use Guidelines*

STEP THREE: DETERMINE PROPER TREATMENT ADDITIVE PROGRAM

Once the proper acid types, volumes, and concentrations have been determined, the acid additive program must be selected. Additives are always required, but it is important that only necessary additives be used. Careful thought should be given to determination of appropriate additives and combinations thereof for each treatment case. There are, arguably, three necessary additives:

- Corrosion inhibitor
- Iron-control agent(s)
- Water-wetting surfactant

Corrosion inhibitor is absolutely, inarguably necessary under all circumstances. There are cases where iron-control additives or surfactants are not absolutely necessary, but for all intents and purposes, they should be included, unless there is a very good reason not to. Additives other than these three should be considered optional. However, more often than not, one or more optional additives are needed, depending on well conditions, formation characteristics and fluid types, and formation damage contributors.

With respect to acid additives, common causes of treatment failure are:

- Additive misuse
- Additive overuse

It is not uncommon to find treatment designs in which unnecessary additives are included. Even worse, unnecessary additives may not only be used, but they may be included in concentrations that are too high. Even necessary additives, when used in excess, can cause more damage than that which originally existed. For example, certain additives, if added in excess, will cause upsets in production facilities, such as foaming or emulsion formation.

Another common problem is the use of multiple additives; two or more may be incompatible with one another. Although service companies have guidelines for additive use, it is not inconceivable that they may not be consulted properly. Consequently, additives may be included that are not compatible. Bear in mind that service company guidelines for certain additive uses are not always readily available to their representatives, especially those located outside the United States and in the more remote locations. As a result, errors in mixing additives can be made. A common error is to mix cationic and anionic surfactants or polymers together. Another error is use of an iron-control agent in an acid where its solubility is low.

It is very important to carefully review the additive program. If you are the customer, you should question those additives not explained in a job proposal, or those you are not familiar with. Additives should have an explainable purpose beyond providing "insurance." That is not sufficient justification.

Other than the straightforward treatment design cases, or where it is otherwise impossible to do so, acid/additive systems should be tested for compatibility with formation fluids. It is not always obvious, especially at extreme conditions, how additives will interact with one another and with formation fluids (oil and water). Service company product and mixing guidelines also should not be overlooked.

Acid additives discussed in this section are:

1. Corrosion inhibitor and inhibitor intensifier
2. Iron-control agent
3. Water-wetting surfactant
4. Mutual solvent
5. Alcohols
6. Nonemulsifier/deemulsifier
7. Antisludging agent
8. Clay stabilizer
9. Fines-fixing agent
10. Foaming agent
11. Calcium sulfate scale inhibitor
12. Friction reducer
13. Acetic acid (additive to HCl-HF)

Some of these additives, especially surfactants, are used for other applications. Also, other additives not listed may exist, especially for very specialized problems or isolated applications. Those listed and discussed are most common and have broad application.

1. Corrosion inhibitor

Corrosion inhibitor is always necessary. It must be added to all acid stages (pickling treatment, acid preflush, main acid, and acid overflushes). I once discussed acid design in a carbonate reservoir with an operations executive for a major oil company. He told me that if concentrated HCl (37% solution) is pumped, corrosion inhibitor is not necessary. His reasoning was that there is not enough water (63% by weight) present in a concentrated acid mixture to cause corrosion. "We used to pump it all the time. It is the 'dilute' acid mixtures, like 15% HCl, that have a lot of water present that are corrosive," he said.

Placing a section of tubing or a corrosion coupon in a beaker filled with uninhibited concentrated HCl acid will confirm that corrosion inhibitor is always required.

Although corrosion inhibitor is always necessary, an excess of corrosion inhibitor can cause problems, such as oil-wetting the formation. Corrosion inhibitors are often cationic polymers that oil-wet sandstones. In general, a corrosion inhibitor concentration greater than 1% may be questionable, unless downhole temperature is greater than 250°F. At higher temperature conditions, inhibitor concentration should generally not exceed 2%, unless corrosion protection requirements are very stringent. Special alloy tubulars, such as chrome tubing, will likely require higher inhibitor loadings for high-temperature protection.

Unfortunately, it is not uncommon for operators to require excessive corrosion protection. I have seen requests for 24–36 hours of protection at 400 °F, for example. This places the service company in a compromising position of accepting a high profit on selling excessive corrosion inhibitor at the risk of causing other treatment problems, which no one desires.

Below 200–250 °F, a corrosion inhibitor in proper concentration (0.1–1%) is usually sufficient to inhibit acid corrosion. At higher temperatures, it may be necessary to include more inhibitor, as well as an inhibitor intensifier or booster. At temperatures of more than 300 °F, an intensifier will be required. Each service company has its own intensifiers or boosters. They are not all the same. Inhibitor intensifiers or boosters often do more to boost treatment cost than corrosion inhibition.

Corrosion tests must be conducted with corrosion coupons representing the metal to be contacted downhole, to properly determine corrosion inhibitor and intensifier loadings. More of either or both is not necessarily better. Optimum combinations exist. Most effective inhibitor intensifiers are in the iodide salt family. Concentrations vary depending on temperature and corrosion inhibitor used.

Service companies also have large databases of corrosion tests under different conditions and with different metals to consult. These should be utilized, especially for high-temperature applications and for wells with unusual alloy steels or special tubulars.

For high-temperature applications, in particular, it is recommended to have laboratory corrosion tests run to confirm that sufficient inhibition will take place. However, it should be understood that only in rare cases is it necessary to have more than 12 hours of corrosion protection. Often, 6–8 hours are plenty. Service companies should not be placed in a position to overload with inhibitor and intensifier in order to meet what amounts to an unreasonable request to provide "insurance."

Table 6–7 summarizes the industry's acceptable corrosion limits.

2. IRON-CONTROL AGENT

Iron control is required in any acidizing treatment. Therefore, an iron-control agent is almost always needed. There are many service company iron-control products and product names. However, products exist in two general categories: (1) iron-complexing or iron-sequestering agents, and (2) iron-reducing agents. One or more of these can be used in an acid mixture. Combinations can be effective, especially at higher temperatures, where dissolved iron contents may be high.

Iron-control agents react with dissolved iron and other dissolved metal ions to inhibit solids precipitation as acid spends and pH increases. Iron-control agents do not reduce the amount of iron dissolved, nor do they

Table 6–7. Corrosion Inhibition—Acceptable 24-hr Limits

Temperature (°F)	Corrosion Limit (lb/ft^2)
<200	.02
200–275	.05
275–300	.09[(a)]

[(a)] For alloy tubulars, should still be .05 lb/ft2 or even less.

Recommended corrosion inhibitor concentration range is 0.1–2.0% (depending on temperature and metal). The recommended inhibitor-intensifier concentration range depends on the type (iodide salts can be 0.2–3%; formic acid, 0.5–5%). Values are for 24 hours of metal/acid contact time.

Table 6-7. *Corrosion Inhibition—Acceptable 24-hr Limits*

reduce or prevent acid reaction with iron compounds. They do prevent reprecipitation of iron compounds by maintaining iron cations in solution.

This may have other benefits, as well, including prevention of sludge formation. Formation of sludge is often enhanced by reprecipitated iron. Therefore, effective iron control may also control, or at least reduce, sludge formation. Iron-reducing agents, such as erythorbic acid, are particularly effective in controlling iron sludging.

Iron control is especially important in tubing pickling treatments. Pickling is designed to remove iron deposits and debris from the tubing prior to acid injection into the formation. Therefore, iron-control agent(s) should be included in the inhibited pickling solution to be pumped down tubing and circulated out prior to the acid treatment.

Iron control is also especially important in the acidizing of injection wells, because iron rust and scale are deposited in the wellbore and at the formation face. Producing well tubulars also typically contain layers of iron oxide or iron sulfide. Acid will dissolve rust very easily and redeposit the iron in the formation if an iron-control agent is not used.

Iron control is also important in formations containing an abundance of iron minerals, such as siderite (iron carbonate) and iron chlorite clay. In the absence of adequate iron-control additive(s), reprecipitation of iron from these minerals may occur in the formation as acid spends and pH increases. Generally, iron dissolved from tubulars presents the greatest problems. However, reprecipitation of iron dissolved from formation minerals can occur and can cause significant formation damage in cases in which high levels of iron are dissolved in the formation itself.

Table 6–8 lists examples of common iron-control agents and concentration ranges in applicable acid mixtures. As can be seen in Table 6–8, some iron-control agents are available in their acid states or as sodium salts. Concentrations would differ depending on whether the iron-control agents are in the free acid form (i.e., EDTA) or one of the sodium salt forms. Service company guidelines must be consulted for proper additive loading. Sodium salt forms preclude use in HF acid mixtures, as spent HF acid will form insoluble precipitates with sodium ions.

In addition to the common iron-control agents listed in Table 6–8, service companies provide other iron-control systems. Citric acid combined with acetic anhydride has been a common system in the past. Special iron-control acid mixtures (blends of acetic, formic, and citric acid) are commercially available.

Newer iron-control agents and combinations thereof have been introduced in recent years. There are also proprietary iron stabilizers, which are typically combinations of iron-control agents and organic acids. Iron reducers from the same family as erythorbic acid are often included in the proprietary additives. The chief merits of the new proprietary blends are increased profit margins for suppliers.

Special iron-reducing agents can be very effective in reducing or preventing acid sludge formation in heavier oil reservoirs and can be considered for special applications. However, the iron-control additives listed in Table 6–8 are the most common and can be used for most applications encountered.

In the treatment of sour wells (wells producing H_2S gas), the possibility of iron sulfide (FeS) precipitation exists. FeS precipitates at a pH of about 2, which poses a potential problem as acid spends and pH rises. Hall and Dill have shown that a mixture of NTA, mutual solvent (ethyleneglycol-monobutylether, EGMBE), and a sulfide modifier can control the precipitation of FeS.[6] A combination with citric and acetic acid is effective as well. Sulfide modifier additives are available from the service companies and may be included in the iron-control additive package.

Type	Temp. Limit	Application [3]
Erythorbic acid	350 °F +	Reducing agent. Use in < 20% HCl; HCl-HF. *10–100 pptg*
Erythorbic acid (Na salt)	350 °F +	Same, except do not use in HF (Na salt). *8–80 pptg*
Citric acid	150–200 °F	Sequestering agent. Can use in all acids. *25–200 pptg*
EDTA (acid form) [1]	350 °F +	HCl and HCl-HF. Limited solubility. *30–60 pptg*
EDTA (disodium salt)	350 °F +	HCl only; do not use in HF acid. *40–80 pptg*
EDTA (tetrasodium salt)	350 °F +	HCl only. Do not use with HF. *50–100 pptg*
NTA (acid form) [2]	350 °F +	All acids. Low solubility in weak acids. *25–350 pptg; 50–100 pptg common*
NTA (trisodium salt)	350 °F +	Same, except do not use in HF (Na salt). *25–350 pptg; 50–100 pptg common*

[1] EDTA is ethylenediaminetetraacetic acid
[2] NTA is nitrilotriacetic acid
[3] pptg is pounds (lbs) per 1000 gallons of acid

Liquid concentrate forms of these iron-control agents also exist.

Table 6-8. *Concentration Ranges of Iron-Control Agents*

3. Water-wetting surfactant

A nonionic water-wetting surfactant is also a necessary acid additive. One should be included in the acid stages. A water-wetting surfactant will aid in the cleanup of acid and will leave the formation water-wet, enhancing flow of oil or gas. For more information on surfactant types, see appendix D.

A water-wetting surfactant should not be added beyond a concentration of 1%. High concentrations of surfactant can cause emulsion and foaming problems in production processing equipment. There are many suitable water-wetting surfactants. Choice is not as important as ensuring that one is included in the proper concentration range.

Sometimes surfactants are provided as "surfactant packages." Surfactant packages may contain two or more surfactants in one product. An example is the combination of a water-wetting or flow-back enhancing surfactant (fluid surface tension reducer) with a nonemulsifier. This is fine; however, one should be careful to make sure that duplication is avoided. For example, a duplication occurs if both a combination surfactant product and a separate nonemulsifier surfactant are included. The nonemulsifier would generally not be necessary unless its need has been established and proven through laboratory fluid compatibility testing.

The recommended water-wetting surfactant concentration range is 0.1%–1.0% (0.1–0.4% is preferable).

4. Mutual solvent

A mutual solvent such as EGMBE can be beneficial.[7] Because mutual solvent is miscible in oil and water, it helps maintain a water-wet formation. Mutual solvent is effective in both oil and gas wells, but more so in oil wells.

EGMBE can be added to acid; however, it may be most effective as an additive to the overflush (ammonium chloride solution). Mutual solvent will strip adsorbed additives, such as corrosion inhibitor, from formation surfaces. Maximum EGMBE concentration is 10%. Concentrations of 3–5% are often adequate. In cases where there are other additive loadings, such as corrosion inhibitor or cationic surfactants, higher concentration of

EGMBE may be called for. In addition to EGMBE, other modified glycol ethers are commercially available from the service companies.

There are also several commercially available mutual solvents and blends of solvents, water-soluble and oil-soluble alcohols, mutual solvents, surfactants, and cosurfactants for acid systems. Such systems are designed to impart strong water-wetting characteristics or to cut through oil and asphaltenic barriers on rock or scale surfaces, increasing acid-rock or acid-scale contact. They are often proprietary and will not be detailed here. All are generally effective. However, it is always best to keep things simple. Therefore, it is suggested that a single mutual solvent, or simple blend, combined with a water-wetting surfactant is suitable for most applications.

There are certain drawbacks to the use of mutual solvents. If used in excess, salts can precipitate from spent acid. Excess mutual solvent also can cause additive separation from acid in the tanks.

The recommended mutual solvent concentration range for EGMBE is 1–10% (3–5% is preferable). Proprietary alcohol blends, if used, vary in recommended concentration ranges.

5. Alcohol

Mixing alcohols, such as methanol or isopropanol, with acid can help unload spent acid in gas well stimulation. This is because alcohol decreases the surface tension of acidizing fluids without adsorbing on the formation like a surfactant. Alcohols react with HCl above 185 °F to produce organic chlorides, which poison refinery catalysts in downstream operations. Therefore, alcohols should only be used as additives in gas well acidizing.

Methanol should be added to acid stages in gas wells that may form gas hydrates downhole. This is a potential problem in deep water completions, where gas cooling in the wellbore may take place, as well as in gas storage wells.

In general, methanol is not safe. Its use should, therefore, be considered only when absolutely necessary. Also, methanol is an inferior mutual solvent. Therefore, it may be prudent to combine it with EGMBE or the like. Isopropanol is an effective mutual solvent on its own. For oil wells, only surfactants, mutual solvent, or gas (N_2 or CO_2) should be used to enhance flowback of spent acid.

The recommended maximum alcohol concentrations are: methanol, 25%; isopropanol, 20%.

6. Nonemulsifier/deemulsifier

Nonemulsifiers are surfactant additives that can be added to acid to prevent acid-oil emulsification. Necessity of a nonemulsifier should be demonstrated with acid/crude oil fluid compatibility tests using produced crude from the well, or zone, to be acidized. Service company labs are equipped to run such testing and have considerable experience doing so. Proper acid treatment planning and additive selection should take advantage of such testing capabilities.

It is important that a nonemulsifier surfactant be either nonionic or anionic in order to keep the rock water-wet. Cationic surfactants oil-wet sandstones. Compatibility with other additives must also be demonstrated. Again, nonemulsifier may already be included in a surfactant package product. In such cases, a separate nonemulsifier may not be required. An excess amount of nonemulsifier will cause emulsions, rather than prevent them.

Deemulsifiers, or demulsifiers, are surfactants that break emulsions already formed. Demulsifiers are usually injected in a nonacid carrier fluid, such as an aromatic solvent. They are not often used, as such treatments are not, and really should not be, required often. Emulsion problems should be addressed during acidizing—through prevention.

The recommended nonemulsifier concentration range is 0.1–2.0% (depends on laboratory testing). The preferred range is 0.1–0.8%.

7. Antisludging agent

A special class of nonemulsifier is the antisludging agent. Acid often reacts with crude oil, particularly low-gravity, high-asphaltenic crudes, to form a sludge. Higher acid concentrations exacerbate the problem. Once formed, a sludge will not dissolve appreciably in the produced oil. Unfortunately, subsequent treatment with aromatic solvent is not efficient, as sludges are quite persistent. In addition, sludges are quite often stabilized by iron, abundantly present in partially or fully spent acid.

Fortunately, certain surfactants are quite effective in preventing the formation of sludge. As mentioned earlier, a proper iron-control agent (reducing agent) also may help to prevent or reduce sludge formation when included in combination with the proper antisludge surfactant.

Acid/crude oil compatibility tests should be conducted to demonstrate the need for an antisludging agent. They also should be used to determine the surfactant and iron-control agent types and concentrations required.

Implementing a crude oil displacement stage in the acid treatment design will substantially reduce the likelihood of sludge formation.

The recommended antisludging agent concentration range is 0.1–1.0% (depends on laboratory testing).

8. Clay stabilizer

A clay stabilizer is often recommended for the purpose of preventing migration and/or swelling of clays following an acid treatment. Common clay stabilizers are either polyquaternary amines (PQA) or polyamines (PA). Polyquaternary amines are considered the most effective. Clay stabilizer seems to be most effective when added to the overflush only. It really is not necessary to add a clay stabilizer to acid, although that is where it is most commonly included.

It would make more sense to inject clay stabilizer with the preflush stage prior to HF injection in order to protect clays during the initial ion exchange. In addition, HF reduces the tendency of clay stabilizer to adsorb on clay surfaces. Therefore, it is considered sufficient and most cost-effective to include clay stabilizer, if it is to be used, in the overflush stage.

The effective clay stabilizer concentration range is 0.1–2.0% (0.1%–0.4% is recommended).

9. Fines-fixing agent (FFA)

In sandstones, most migratable siliceous fines are actually not clays.[8] The problem of fines migration in sandstones has been studied extensively, led by Sharma and Fogler, in particular.[9–11] Clay stabilizers do not control migration of nonclay fines such as quartz, feldspar, and mica.

There is a commercial additive that effectively prevents clay migration and /or swelling, and the migration of nonclay siliceous fines.[12, 13] The fines-fixing agent (FFA) is an organosilane, which reacts in situ to form a poly-siloxane of varying lengths, binding siliceous formation fines in place. The formation fines are thereby stabilized, preventing migration and subsequent plugging at pore throats. FFA may be used in acidizing treatments, or in nonacid treatments in a brine carrier fluid.

FFA/acid treatments have been pumped successfully in the Gulf of Mexico as well as in California, where siliceous fines migration problems are prevalent. FFA additives have been used most commonly as an additive in the ammonium chloride overflush, particularly in treating gravel pack completions in which fines migration is a problem.[14] The silane FFA may also be combined with conventional clay stabilizer (e.g., 0.5% FFA, 0.1% clay stabilizer) to ensure stabilization of both clay and nonclay siliceous fines. FFA may also be used as an additive to acid mixtures, including HF, imparting both fines-stabilization and retardation to the HF-rock reaction.[13]

The recommended fines-fixing agent (FFA) concentration is 0.5%–1.0%.

10. Foaming agent

The use of nitrogen and a foaming agent can assist return production of spent acid in gas wells by reducing fluid gravity and surface tension of the fluids injected. Nitrogen alone [300–1500 standard cubic feet (scf)/bbl of injected solution] can also improve treatment flowback and accelerate return to production. Acid containing both a foaming agent in low concentration and nitrogen continuously commingled is used in gas well stimulation. Inasmuch as foam is an acid placement technique, acid may also be foamed to improve placement.

It is believed that the higher the foam quality (65–80%), the better the placement, or diversion, of acid. However, this relationship is not necessarily true. Optimum foam quality is a function of permeability, porosity, and reservoir fluid properties.

The recommended foaming agent concentration range is 0.3–0.8%.

11. Calcium sulfate scale inhibitor

It may be advisable to include a calcium sulfate ($CaSO_4$) scale inhibitor in the acid stages or the overflush if treating a well containing high sulfate ion concentration (>1000 ppm) in the formation water. $CaSO_4$ scale inhibitors are typically phosphoric acid or polyacrylate polymers. There are also certain proprietary acid systems containing acids that may inhibit sulfate scale formation. Scale inhibitor chemical compatibility must always be demonstrated.

Calcium carbonate ($CaCO_3$) scale inhibitor squeeze treatments may be accomplished apart from acidizing. They may be recommended following acid treatment or acid cleanup of the formation. Unlike calcium sulfate scale inhibitors, calcium carbonate inhibitors are not compatible as additives to acid treatment stages.

The recommended $CaSO_4$ inhibitor concentration range varies depending on type and severity of scale potential.

12. Friction reducer

Pipe friction can increase acid treatment injection pressures, thereby lowering injection rates, which may be undesirable. This is especially true in smaller diameter injection strings, such as coiled tubing. Friction pressure represents increased pumping energy resulting from fluid drag on pipe. Long chain polymer (e.g., polyacrylamide) gelling agents are used to make HCl more viscous in carbonate acidizing treatments.

These gelling agents also act as friction reducers when used in lower concentrations and may be added to any acid. In effect, friction reducer "dampens" fluid movement in turbulent flow, thereby reducing friction drag (and subsequently injection pressure) closer to laminar flow regime.

Friction reducer should be used in deep well treatments and high-rate treatments. Service companies have friction pressure plots or tables for calculating the excess injection pressure due to friction for various fluids, including solvent, brine, and certain acid mixtures.

The recommended friction reducer concentration is 0.1–0.3%.

13. Acetic acid

Adding acetic acid to the HCl-HF stage is an option that may help in reducing precipitation of certain aluminosilicates as the pH of the HCl-HF mixture rises with acid spending. A small amount of acetic acid delays the precipitation of aluminosilicates by buffering the acid mixture (maintaining a sufficiently low pH) and by complexing with aluminum (chelating effect).[5]

The recommended concentration is 3%.

There are many additive applications and choices. The task of proper additive selection can become confusing and is seemingly overwhelming at times. The natural aversion to chemistry often discourages the engineer from exploring additive use and recommendations as far as he or she should.

It is best to consider additives by:

- Need
- Classification or type

Need must be based on existing damage present and on damage that can potentially be caused by the acid treatment. Once the absolute additive needs are identified, additive types can be considered. Selecting by type at least limits choices to about 12 categories or so. Within each type or classification, there are many choices, with varying effectiveness and cost.

Regardless of affiliation—oil company or service company—one should ask questions of the appropriate service company technical experts. One also should consult product and application information to understand recommended products used. Hopefully, selections then will be limited to only those additives that are absolutely required.

Service companies possess excellent information on products. However, this information is often limited to in-house access, and even then it is difficult to locate in some cases. In one way or another, though, especially through questions and investigation, one may gain the understanding and confidence necessary to make proper additive selections and economic choices. This helps one to properly evaluate treatment proposals for each acid treatment case.

STEP FOUR: DETERMINE TREATMENT PLACEMENT METHOD

Determination of the proper fluid placement method is a key factor in acid treatment design in both carbonates and sandstones. Treatment success can hinge on it. More often than not, some method of placing or diverting acid is required to distribute acid across the zone or zones of interest. This is especially true in matrix acidizing.

Selection of an acid placement method is an area sometimes requiring creativity and always requiring openness to options and combinations of such. With respect to selection of acid placement methods, charts and service company and operator guidelines have proven repeatedly to be of limited usefulness.

The inclination we engineers have to develop and use standardized selection methods for certain sets of well and treatment conditions is not appropriate for acid treatment placement and diversion. Therefore, only limited information in that regard will be provided here. You are instead encouraged to draw upon your creative and cooperative talents to develop an understanding and facility with acid placement design. It is important to remember, however, that no method or combination of methods is going to be effective in all cases.

In all cases, however, service company and operator representatives should discuss and work out a mutually acceptable method of placement and/or diversion to be included in an acid treatment design procedure. It may be based on prior experiences and examples. It is important, though, to treat each well treatment case as a new, stand-alone treatment. With that said, the remainder of the chapter is intended to provide a basis for placement/diversion selection and development.

The importance of treatment placement was evident and recognized in the earliest acid treatments. In his patent, Frasch noted the need for a rubber packer for isolation so that acid could be selectively injected into the formation. Although this need has been recognized from day one, the absence of proper acid placement is probably still the biggest reason acid jobs fail (besides improper well diagnosis). A well-conceived, properly designed treatment in all other aspects (formation damage assessment, selection of acid types, concentrations, volumes, and additives) can go for naught if the treatment is not properly placed. The zone of interest must be sufficiently contacted by stimulation fluids.

Outside of short, homogeneous formation intervals, which are rare, perfect "coverage" of the treatment zone is not often possible. Poor zone coverage during acidizing is often the norm. It is the result of treatment interval heterogeneities, such as:

- Varying permeabilities in different interval sections or zones
- Varying degrees of formation damage
- Varying reactivities to acid
- Varying formation pressures
- Varying fluid viscosities
- Presence of natural fractures
- Combinations of the above

When encountering very long treatment zones, one or more of these heterogeneities is likely to exist. Most are a result of the sheer height or length of the zone(s) to be treated, which can be thousands of feet, especially in horizontal well completions. Even if zones to be treated are of manageable height, they may contain varying permeability sections or streaks, as well as portions of differing severity of damage within perforations and the formation. Under such circumstances, uniform acid distribution is impossible.

Naturally, during injection, acid takes the path of least resistance, as does any injected fluid. Anyone who has worked on acid treatment design or treatment execution has no doubt heard that before, or said it before. The path of least resistance will be those formation sections or layers with the highest permeability and least damage. Higher permeability streaks, or natural fractures, will accept fluid more readily than zones of lower permeability or injection capacity.

Furthermore, zones with lower formation or reservoir pressure offer less resistance to injection, as do those containing the lowest viscosity fluids (for example, gas versus oil). Further encouraging the flow of acid to the limited paths of least resistance is the reactivity of acid with the formation. The injection capacities of the paths of least resistance will increase even more as the acid treatment progresses.

To approach full damage removal, acid must be "diverted" to the sections that accept acid the least—those that are most damaged. While it may be difficult to imagine achieving perfect zone coverage during acidizing, fortunately improvement in zone coverage may still go a long way in improving stimulation response. This is true especially in severely damaged formations. Therefore, an attempt to modify the acid treatment injection profile is desirable and should be made with such means as are available.

To achieve or approach uniform acid distribution, the acid injection rate per unit area to be treated must be varied. Acid injection rate per unit area can be described by Darcy's equation, as follows:

$$Q/A = k\Delta P/L\mu$$

where

Q is the injection rate
A is the surface area of zone
k is the permeability
ΔP is the differential injection pressure
L is the zone height
μ is the fluid viscosity

The goal in acid placement is to equalize the acid injection rate per unit area (Q/A) across the entire treatment interval, divided in "n" sections:

$$K_1\Delta P_1/L_1\mu_1 = K_2\Delta P_2/L_2\mu_2 = K_3\Delta P_3/L_3\mu_3 = \ldots K_n\Delta P_n/L_n\mu_n$$

For many years now, there have been and still remain two basic methods for placing acid. They are:

- Mechanical placement
- Chemical diversion

Chemical diversion is a method for approaching equalization of injection rate per unit area. Mechanical placement alone is more of a method to allow acid to at least contact most of the zone of interest, or to impart more uniform distribution of acid across the treatment interval. Combinations of the two can be more effective than either one alone and are often employed.

MECHANICAL PLACEMENT

Mechanical placement of acid was the first means thought of for improving contact of the interval to be treated. Currently, the following general acid placement methods exist:

- Packer systems
- Ball sealers
- Coiled tubing

They are not necessarily stand-alone methods, as coiled tubing injection can be combined with packer isolation, as can ball sealers.

Packer systems. Mechanical placement is the most certain method of acid diversion. Mechanical placement or diversion includes the use of inflatable straddle packer systems, bridge plug and packer combinations, and special wash tool packer systems. These are used to physically isolate shorter sections of longer treatment zones.

Mechanical placement with isolation packer systems allows for the injection of acid within a limited treatment interval at a time. Other treatment intervals within the total zone can be treated individually in stages. Figure 6–1 is a depiction of treatment of a limited interval with coiled tubing through a packer setting in a horizontal well section.

There are a number of commercial packers and packer system manufacturers. Improvements in such treatment tool systems and extensions of their application are continually made. Information on packer systems and their uses, including design for specific applications, can be obtained from the major packer manufacturers, such as Baker Oil Tools, Weatherford, and TAM International.

Isolation packers for use in acid stimulation are only effective in a well completion if acid will enter perforations and continue flow into the formation. If a poor cement job is present, acid may channel vertically along the

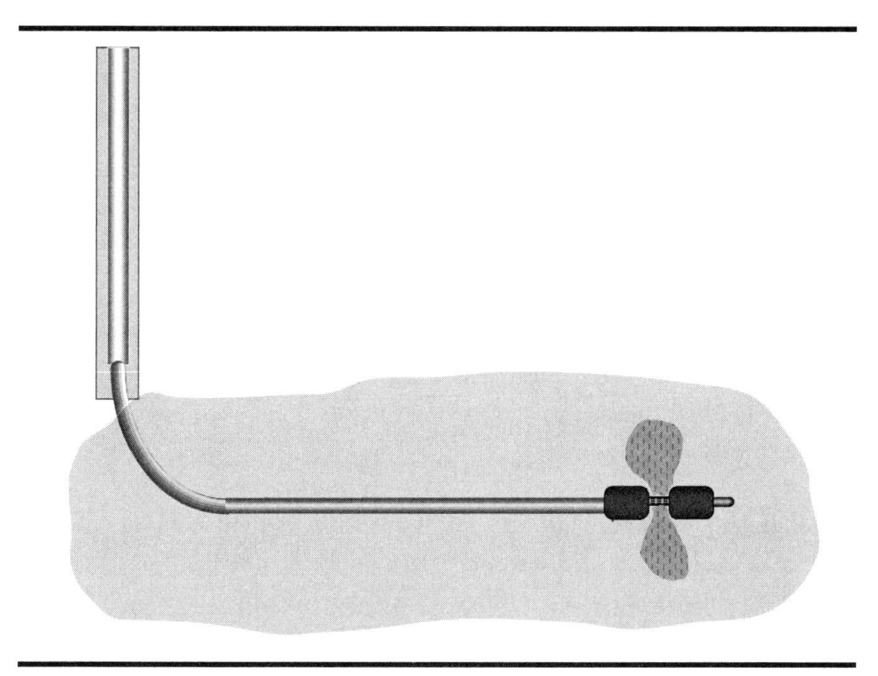

Fig. 6-1. *Treatment of a limited interval with coiled tubing through a packer setting in a horizontal well section*

cement/formation interface, defeating the purpose of mechanical placement. Of course, mechanical placement of acid with packer systems is only valid in perforated, cemented pipe completions. As of this writing, reliable mechanical placement techniques for small diameter, monobore (slimhole) completions do not exist.

Packer systems for acid placement in horizontal wells do exist, however, and can be effective. The limitation that does exist with such systems is the number of placement settings or injection stages that can be applied. Typically, the rubber packer sealing elements may not withstand more than perhaps 8–12 settings in a particular well condition before the risk of packer failure becomes high. This number depends on the packer and downhole conditions.

However, the point is that in extremely long horizontal completions popularized today, mechanical placement may be of limited value, if not totally ineffective. This is because the isolated interval length per stage may have to be too long for all practical purposes as a result of the limit on settings (deflation/inflation cycles) that can be withstood. In such cases, it is sometimes possible, and even desirable, to combine mechanical placement with chemical diversion methods.

Another limitation to mechanical placement is cost. Packer systems often require a rig. The specialty packers themselves may be expensive, as well. Nevertheless, mechanical placement with packers is the most reliable, and it is the best method for improving treatment zone coverage during an acid stimulation treatment.

Ball sealers. A popular mechanical diversion method is the use of "ball sealers." Ball sealers are not an entirely reliable method of diversion. However, under the right conditions, they may provide the most effective diversion of any of the options available. Ball sealers were originally introduced in the mid-1950s. They are just what the name implies—balls that are pumped in an acidizing treatment, intended to seat on perforations to create a temporary seal. The acid is thereby diverted to other perforations as the treatment progresses.

Ball sealers are added to treatment fluids with special equipment (a "ball gun") when diversion is needed. Balls are removed from perforations at the end of a treatment once injection is terminated and pressure in the wellbore drops, allowing the ball sealers to fall out of perforations. Balls either flow back during production and are collected, or drop into the well rat hole.

Ball sealers are most effective in newer wells with a limited number of perforations. In older wells, with damaged or compromised perforations, or with a large perforation density (>4 shots per foot), ball sealer effectiveness is reduced. Also, ball sealers are only effective when casing is well cemented and no vertical channeling will occur behind the pipe during acid injection. Similarly, channels from the perforations into the formation, such as conductive natural fractures, will reduce the effectiveness of ball sealers.

Furthermore, the nature of the perforations has an effect on ball sealing efficiency. The more smooth and symmetrical a perforation is, the better the ball sealer will seat and create diversion—or the better the "ball action," as it is called. The more irregular a perforation, the less chance it has of being adequately sealed. Unfortunately, the quality of perforations in this regard is not often known for sure.

Ball sealer diameters typically range from 5/8″ to 1 1/4″, although probably more than 90% of balls pumped today are 7/8″ diameter. Newer ball guns are designed to deliver the 7/8″ diameter balls, which have the broadest application with respect to perforation diameters and injection tubing sizes. The rule of thumb for selecting ball size to achieve an adequate seal is that the ball diameter should be about 1.25 times the perforation diameter.

Other factors to consider in the use of ball sealers are:

- Injection string diameter
- Acid injection rate
- Ball density

Injection string diameter should be at least 3 times the ball sealer diameter. Therefore, it is best to be pumping an acid treatment through tubing at least 2 7/8″ diameter.

Effective ball sealing action is a function of injection rate. There are rules of thumb regarding rate, such as minimum required rate for ball action is 1/10 to 1/4 bbl/min/perforation. However, it seems that an injection rate of 2 bbl/min, at the very least, is needed to achieve sufficient ball action. The higher the rate, the better.

The injection rate minimum also depends on the ball density relative to the acid. Ball densities typically vary from about 0.9 specific gravity (s.g.) to about 1.4. Ball sealers with densities less than 1.0 are called buoyant ball sealers or floaters. Newer, conventional ball sealer products are of the floater or neutral density variety. The concept was developed at Exxon.[15] Ball sealers with specific gravities greater than water or acid (1.1+) are called sinkers for obvious reasons. Older, conventional ball sealers are of the sinker variety.

Floaters or neutral density ball sealers are designed to be produced back following acidizing. Sinkers are designed to fall to the rat hole following treatment. They are not expected to be recovered. Of course, sufficient rat hole must exist to consider sinkers that will not be produced back. Sinkers require higher injection rates to overcome the drag force of the balls in the lighter fluid.

An important consideration in selecting the proper ball sealer is the settling velocity of the ball in the carrying fluid. The settling velocity of a ball sealer in a Newtonian fluid can be defined as follows:

$$V_s = [4g_c D(p_B - p_F)/3f_D\, p_F]^{1/2}$$

where

V_s	is the settling velocity, ft/sec
D	is the diameter of the ball (ft)
p_B	is the specific gravity of the ball
p_F	is the specific gravity of the transport fluid
g_c	is the gravity acceleration (use 32.2 ft/sec)
f_D	is the friction drag coefficient (essentially a constant equal to 0.44)

Settling rate in 15% HCl, for example, can vary from about 24 ft/min for a 1.1 s.g. ball to almost 90 ft/min for a 1.4 s.g. ball. For buoyant ball sealers, or floaters, the settling rate will be negative—meaning the ball will rise, rather than sink or settle. In 15% HCl, a 0.9 s.g. ball will have a rise rate of about 65 ft/min.

In 28% HCl, a 0.9 s.g. ball will have a rise rate of about 74 ft/min. These rise rates can usually be easily overcome. For example, in 2 7/8″ tubing, the velocity of acid movement at 1 bbl/min is 173 ft/min (5.615 ft^3/min divided by pipe cross-sectional area). In larger tubing, acid velocity decreases, so higher injection rates are required.

To overcome settling rate, it is recommended to pump an excess of ball sealers. With sinkers, rather than pumping 1 ball per perforation, it is recommended to pump up to 200% excess. With floaters, it is recommended to pump about 50% excess, because the drag force during injection may not be high enough to seat balls with full efficiency. However, because of their lower density (less propensity to drop), a smaller excess is required.

Ball sealers are available in different materials, most commonly rubber-type. RCN (rubber-coated nylon) is most common. Selection depends on density and temperature limit requirements.

Relatively new on the market are the biodegradable ball sealers.[16] Biodegradable ball sealers are made of animal protein material that degrades at certain temperatures. With biodegradable ball sealers, there is no issue of plugging, recovery, or loss to rat hole. These conveniences come at a higher price. However, it may be worth it, especially in more complex completions or those containing permanent downhole tools. The practical temperature limit of biodegradable ball sealers is probably about 200°F. However, the temperature limitation may be exceeded in very short treatments (i.e., short pump times).

Coiled tubing. Coiled tubing (CT) is a very useful tool for improving acid placement. Great strides in coiled tubing technology have been made. Coiled tubing utility is versatile. In fact, stimulation treatment placement is probably not one of the best coiled tubing applications, generally speaking. It must be considered carefully. However, when applied properly, it is an excellent tool.

Coils now exist in many sizes and maximum allowable depth ratings, from 1″ diameter to 3″ diameter. Most common coiled tubing strings used in acidizing and wellbore cleanouts are the 1 1/4″ and 1 1/2″ strings. Larger, higher rate treatments utilize 1 3/4″, 2″, 2 3/8″, and 2 7/8″ coils. The larger, 3 1/2″ strings are not often used in acidizing, at least as of this writing.

Coiled tubing is of less use in fracture acidizing because of rate limitation. It is still best to pump fracturing treatments through larger strings, such as production tubing. Coiled tubing is most useful in matrix and wellbore treatment.

Coiled tubing offers some major advantages in acidizing, including:

1. Ease with which an acid injection can be terminated, if it appears that continuing injection is not doing any further good, and switched to flush. The total volume in the CT string is small and can be quickly displaced.

2. Ease with which treatment displacement with nitrogen can be achieved quickly to push reactive fluids away from the wellbore—energizing the near-wellbore fluid zone, thereby enhancing flowback.

Disadvantages include:

1. Pump rate limitations. Smaller diameters cause higher friction pressures, which may limit treatment injection rates to lower-than-desirable levels. Acidizing through production tubing, or drill pipe, for example, will allow higher rates.

2. If solids are needed (perhaps for diversion), there may be problems pumping them through smaller diameter CT strings.

3. Acid mixtures must be very thoroughly mixed and must remain that way prior to and during injection. Corrosion in a CT string is especially disastrous. Small pinholes or pits within the string can quickly lead to tubing failure and a major fishing workover—not to mention the safety aspect of such a failure.

Overall, coiled tubing is very effective in placing acid, especially in smaller treatments, and treatments for damage very near the wellbore. Coiled tubing is also a preferred placement technique in wellbore and near-wellbore treatment in long horizontal wells. Coiled tubing can be used to spot fluid along the zone, while drawing or reciprocating the tubing along the perforated or open-hole interval of interest.

An acid treatment in a vertical well also may be injected while coiled tubing is reciprocated along the treatment zone. Combining coiled tubing with straddle packer assemblies for selective zone placement is a standard method for improving zone coverage. Special nozzles or jetting tools can be attached to the end of the coil to enhance contact of damage with acid.

Coiled tubing is a natural medium for injection of foamed acid or foam diversion. The smaller diameter coil allows for maintenance of foam quality and stability during injection. This increases the chance of achieving diversion with foam, which is not often efficient or even possible.

Chemical diversion. The first attempts at acid treatment placement used chemical "diverting" additives, according to Harrison.[17] In 1936, a soap solution that reacted with calcium chloride to form a water-insoluble, oil-soluble calcium soap was used in a HCl acid treatment. This diversion idea led to the development, or discovery, of more sophisticated chemical diversion methods. These included $CaCl_2$ solution (heavier than acid), cellophane flakes in aqueous solution, and oil-external emulsions.

In the mid-1950s, naphthalene flakes were introduced. Naphthalene is oil soluble and sublimes at about 175°F, giving it application in oil wells and gas wells of moderate temperature. At about the same time, other diverters were introduced, as well. These included crushed limestone (for HCl treatments only), oyster shells, sodium tetraborate, perlite, gilsonite, paraformaldehyde, and chicken feed. Success was mixed with all of these.

These diverters were eventually replaced almost entirely by rock salt (NaCl). While rock salt has continued to be a very effective diverting material, it is only applicable in non-HF treatments. As HF acidizing became popular again in the 1960s, the need for suitable chemical diverting agents spurred the development of completely soluble materials, such as wax/polymer blends, hydrocarbon resins, and benzoic acid. Oil-soluble resins, including a fine particle form for gravel pack completions, became quite popular in the 1980s. However, their popularity has waned in preference of benzoic acid particulates and gels and foam diverters for HF acid treatment diversion.

Today, benzoic acid (flakes or fines) is probably the most broadly applicable diverter, as it is soluble in water and oil and sublimes in gas. Benzoic acid may also be dissolved in water (as ammonium benzoate) and in alcohol. When contacted by water at reservoir pH, benzoic acid dissolved in either aqueous or alcohol solution is precipitated as a smaller, softer, more easily deformed particle. This form of benzoic acid diverter may be used with more confidence in gravel pack completions, for example.

Particle size may be adjusted with surfactant added. Particulates, such as benzoic acid, may be pumped either in slugs or continuously as an additive to acid mixtures. Adding continuously may be preferable at higher injection rates (e.g., 2–5 bbl/min).[18] Benzoic acid is limited to use with temperatures up to about 250°F.

Deformable wax beads are also useful, especially in acidizing naturally fractured formations and carbonates.[19] They are limited by temperature (melting point). The range of use can be extended with the use of pretreatment cool-down pad, when possible. Wax beads are now being used in horizontal well stimulation. Bottomhole temperature should be 25–35 °F higher than the melting point of the wax beads used. However, the treating temperature should be less than the melting point, of course, to allow for temporary diversion required.

Oil-soluble resin is not as popular as it once was; however, it still has occasional application.[20] Oil-soluble resin application is limited because the melting point is more than 300 °F. Therefore, removal must be entirely from dissolution in produced oil. If that is not accomplished, a separate solvent treatment must be pumped to remove the diverter. Other degradable diverters, such as forms of benzoic acid and wax beads, are therefore preferable because of their lower melting points (see Table 6–9).

Rock salt is still the chemical diverter of choice for routine HCl and non-HF treatments. Graded rock salt is the current embodiment. Combined particulates, such as graded rock salt and benzoic acid flakes, rock salt and wax beads, or benzoic acid and wax beads can be effective. They can provide the benefits of a harder material and a softer, more malleable one, enhancing the temporary block created in perforations.

Of broad use these days, as alternatives to particulate diverters alone, are foam and gel diverters.[21, 22] Foams do not always work well. However, in certain areas, foam diversion is effective. It is useful in gravel pack completions, for example, where particulates do not pass well through the pack or through screen slots. Generally, foam is more effective in higher permeability formations with deeper damage. Also, foam diversion is probably most useful in gas well acidizing.

Gels provide a more reliable method of diversion than foam, at the higher risk of damage potential. To reduce damage, gel concentration must be limited. Lower concentration HEC (hydroxyethylcellulose) gels (pills or slugs, 10–20 lbs/1000 gal) exhibit viscosity-controlled leak-off, which can provide sufficient diversion. However, if the gel diverter does not break and clean up completely, perforations can be damaged. A separate treatment must be conducted to remove the gel. This would be highly undesirable for any operator.

In any case, properly mixed gels, foam gels, or gels containing particulate diverters are useful in oil and gas wells in which permeability is relatively high and a more "stout" diverter is needed.

Type	Applications	Concentration[†]
Rock Salt (MP[‡]: 1472°F)	HCl and non-HF treatments	Perfs: 0.5–2 lbs/ft Open hole: 5 lbs/ft^2 formation
Benzoic acid flakes (MP: 252°F)	Gas, oil, injection wells	Perfs: 0.5–1 lb/ft Open hole: 2.5 lbs/ft^2 formation
Wax particles (beads) (MP: 150–160°F)	HCl and HF treatments; Do not use in gas wells.	Perfs: 0.25–1 lbs/ft Open hole: 2.5 lbs/ft^2 formation
Oil-soluble resin (OSR) (MP: 328°F)	HCl and HF treatments; Do not use in gas wells or brine injection wells.	Perfs: 0.5–2.5 gal fluid containing 0.5% OSR per perf. Open hole: 5–20 gpt
Foam	Preferably in gas wells, higher permeabilities; Can gel or emulsify for added viscosity.	60–80 quality
Ball Sealers	Sinkers Neutral density or floaters	200% excess 50% excess

[†]Diverters are in solution at same concentrations per gallon fluid
[‡]MP is melting point.

Table 6-9. *Preferred Diverter Guidelines*

In the end, and in all cases, no diverter is fool-proof. All are still hit-or-miss. However, best efforts can be attempted with careful consideration, investigation, and discussion for any well treatment case. In each well case, the principal service company representative and operator representative responsible for treatment design should work up the diverter/placement method and procedure. These should meet objectives to the mutual satisfaction of both parties.

Perhaps the greatest need in the acidizing industry is for a very effective, nondamaging, inexpensive diverter or acid placement method. Inasmuch as treatment conditions are so different from one well to another, such a diverter is probably a fantasy. However, it is possible to continue improving upon existing methods and certainly improving on well assessment and selection of acid placement technique. Table 6–9 lists currently used commercially available chemical diverter types and widely accepted industry guidelines for use.

STEP FIVE: ENSURE PROPER TREATMENT EXECUTION AND QUALITY CONTROL

In the early 1980s, King and Holman of the Amoco Production Company produced a booklet entitled "Acidizing Quality Control at the Wellsite." [23] The booklet was an Amoco in-house set of guidelines. However, the booklet has become, in one form or another, part of in-house guidelines of other companies. Many copies of the little red book have found their way outside and beyond the walls of Amoco and into the hands of service company and other oil company technical personnel.

Because of its utility and immeasurable value and importance, many, including myself, have benefited from its use. Years ago as I was present onsite to observe an acid job, I unexpectedly received my copy from a well-meaning service company operations supervisor. I had no choice but to accept it happily, as it was in the mutual interest of improving acidizing treatment execution.

This section lists the steps that should be taken toward quality assurance. Inasmuch as quality control applies to all acid treatments, these steps are revisited and further detailed in chapter 16. Safety at the job site is discussed in chapter 17.

Quality control steps to be implemented in the execution of a sandstone acidizing treatment, as recommended by King and Holman (with slight modification), are as follows:

1. **Quality control during rig-up of equipment:**

   A. *Inspect all tanks that will be used to hold acid or water. The tanks must be clean. Small amounts of dirt, mud, or other debris can easily ruin an acid job.*

   B. *Make sure the service company has the equipment to circulate the acid tank prior to pumping.*

   C. *The line to the pit or tank should be laid and ready to connect to the wellhead so the acid can be backflowed immediately after the end of the overflush.*

2. **Quality control before pumping:**

   A. *Check service company ticket to be sure all additives for the job are on location.*

   B. *Circulate the acid storage tank(s) just before the acid is injected into the well.*

   C. *Check the concentration of acids with a test kit.*

   D. *Make sure the service company personnel know the maximum surface pressure and stay below that pressure.*

   E. *Check the pressure-time recorder, or any other on-site treatment monitoring or evaluation mechanism for proper operation.*

   F. *Acid-clean (pickle) the tubing.*

3. **Quality control during pumping:**

   A. *Control injection rate. Maintain surface annulus pressure at or below 500 psi during treatment.*

B. *Watch the pressure response when acid reaches the formation.*

The surface pressure should slowly decrease if the rate is held constant, indicating skin removal. If the surface pressure rises sharply or rises continuously for several barrels of acid, the acid may not be removing the damage or may be damaging the formation.

C. *Note the pressure response when the diverting agent reaches the formation.*

The surface pressure should rise slightly. If there is no response, more diverter or a different diverter may be needed in the future.

D. *Never exceed the breakdown pressure of the formation in a sandstone acidizing treatment, unless absolutely necessary (but only temporarily to initiate injectivity).*

E. *In the final flush, make sure acid is displaced from the wellbore.*

4. Quality control after pumping /during flowback

A. *Do not shut the well in after acid injection. Flow the well back to the tank or pit as soon as the flow line is connected.*

B. *Collect at least three 1-quart samples of backflowed acid for analysis. Sample the acid backflow at the beginning, middle, and near the end of the flow. If on swab, get a sample from every other swab run.*

If possible, a complete analysis of acid returns can be made by the service company using spectroscopic methods.

C. *Take the treatment report and the pressure charts to the office for evaluation and placement into the well file.*

The stimulation pressure record can be a very valuable tool in evaluating acid effect on the formation.[23]

Again, these quality control steps are presented with more discussion in chapter 16, as they apply to acid treatments in general.

STEP SIX: EVALUATE THE TREATMENT

Treatment evaluation involves the following:

- Pressure monitoring during injection
- Flowback sample analysis
- Production rate comparison and analysis
- Well test analysis (skin removal)
- Payout and ROI (return on investment)

PRESSURE MONITORING DURING INJECTION

Monitoring the pressure during injection will indicate diverter effectiveness and possibly evolution of skin removal. If pressures are unchanged, it may be that the acid is not doing any good. If pressures continue to increase at constant or declining injection rates, the treatment may have been causing damage rather than removing it.

There are methods for evaluating pressure responses during acidizing. They are based on interpreting recorded wellhead pressure values (or the subsequent derived bottomhole pressure values) and corresponding injection rates as treatment progresses. This is done in order to derive the evolution (hopefully reduction) of the skin factor during matrix acidizing. Methods have been presented by:

- McLeod and Coulter[24]
- Paccaloni[25, 26]
- Prouvost and Economides[27]
- Behenna[28]
- Hill and Zhu[29]

Prouvost and Economides and most recently, Montgomery and Economides, provide descriptions of these methods, which will suffice.[2, 30]

Each method has uses, depending on conditions and the well and reservoir information available. These methods (especially the first three) and the modifications and alternatives that have followed are considered breakthroughs in matrix acidizing evaluation, and they merit study.

Service companies are equipped to provide on-site, real-time evaluation of acidizing treatments. However, these techniques require accurate data, and therefore, their usefulness is often limited. These limitations are discussed elsewhere, as well.[30] At the very least, constant injection rates and an accurate pressure-time recorder are required. Despite limitations, it is recommended that in fields where multiple acidizing treatments are planned, real-time evaluation be used, if possible. It is most important to gather such data for initial treatments in a field. Future success may depend on it.

Montgomery *et al.* have described techniques that can be used to apply real-time treatment evaluation to provide diagnostics and interpretation, enabling proper decision-making and use of these methods.[31]

FLOWBACK SAMPLE ANALYSIS

Analysis of flowback samples is important for observation of sludge, emulsion, solids production, and related problems. Spent acid compositional analysis (see chapter 16) can shed light on problems, especially acid reaction product precipitation, and can help in optimizing future treatment design.

PRODUCTION RATE COMPARISON

Comparison of production rates (before and after) is the most obvious and simple measure of success. Rate comparison should only be made seriously after all spent acid has been returned and well production has returned to formation fluids only. Changes in oil to water or gas to water ratios are important to monitor. Acidizing should not preferentially stimulate a water zone. If that happens, it may be that the reservoir was not understood, or that diversion was not effective.

Poststimulation production data should be carefully compared to prestimulation data to make sure that reported production rates are under comparable before-and-after conditions. There are many cases where certain factors are not noticed, or are misread, leading to a gross misinterpretation.

Opening a surface choke (increasing drawdown), opening gas lift valves downhole, changing a pump or surface valve, etc., all can lead to higher production rates. This data can be misinterpreted or even intentionally misreported as increases due to stimulation.

POST-STIMULATION WELL TESTING

Post-stimulation well testing is the truest indicator of success or failure. Well testing should be conducted before and after stimulation to identify pretreatment and posttreatment skin values (preferably identifying damage skin, s_d). If this is accomplished, a genuine before-and-after comparison can be made, in conjunction with prestimulation and post-stimulation production rates.

PAYOUT AND COST RETURN ON INVESTMENT

Evaluation of payout and cost return on investment are among the bottom-line factors to the operator. They represent the type of indicators that the service companies thrive on, and, seriously, should hear about. The service companies can be the most ambitious messengers of success, provided it is to their benefit, too.

Payout and return on investment (ROI) are simple to calculate, depending on the fluid(s) produced (gas or oil), the treatment costs, and the day-to-day well operating costs. Payouts on the order of a day or two to several weeks are almost always more than acceptable. There are those who would wish to replace acidizing completely with fracturing or frac-and-pack treatments. However, acidizing is unmatched in its ability to generate immediate cash flow, rapid pay-out, and high return relative to cost.

Daneshy shows that production enhancement treatments can be very effective in reducing the cost per barrel equivalent (oil or gas), as well as generating substantial financial benefit.[32] (These treatments would include matrix acidizing in damaged sandstone formations.) He suggests that the very low cost/reward ratio of such treatments could justify a more aggressive approach that might involve greater risk.

Relatively low-cost formation damage removal treatments, such as acidizing, are unrivaled in their potential financial significance. This is especially true in medium to high permeability reservoirs, not often ideal candidates for hydraulic fracturing.[32]

REFERENCES

1. H. O. McLeod, "Matrix Acidizing," *Journal of Petroleum Technology* (Dec. 1984): 2055–69.

2. M. J. Economides and K. G. Nolte, editors, *Reservoir Stimulation,* 2d ed. (Schlumberger Educational Services, 1989), chapter 16.

3. W. J. Lee, *Well Testing,* Society of Petroleum Engineers Textbook Series Vol. 1 (Society of Petroleum Engineers, 1982): 5.

4. H. Perthius and R. Thomas, "Fluid Selection Guide for Matrix Treatments," 3d ed. (Tulsa, OK: Dowell Schlumberger, 1991).

5. C. E. Shuchart and S. A. Ali, "Identification of Aluminum Scale With the Aid of Synthetically Produced Basic Aluminum Fluoride Complexes," *SPE Production & Facilities* (Nov. 1993): 291.

6. B. E. Hall and W. R. Dill, "Iron Control Additives for Limestone and Sandstone Acidizing of Sweet and Sour Wells" (paper SPE 17159, Society of Petroleum Engineers, 1988).

7. J. L. Gidley, "Stimulation of Sandstone Formations with the Acid-Mutual Solvent Method," *Journal of Petroleum Technology* (May 1971): 551–58.

8. T. W. Muecke, "Formation Fines and Factors Controlling Their Movement In Porous Media," *Journal of Petroleum Technology* (Feb. 1979): 144–50.

9. M. M. Sharma, "Fines Migration in Porous Media," *American Institute of Chemical Engineers Journal* (Oct. 1987): 1654–62.

10. C. Gruesbeck and R. E. Collins, "Entrainment and Deposition of Fine Particles in Porous Media" (paper SPE 8430, presented at the Society of Petroleum Engineers Annual Technical Conference and Exhibition, Las Vegas, NV, Sept. 23–26, 1979).

11. G. A. Gabriel and G. R. Inamdor, "An Experimental Investigation of Fines Migration in Porous Media" (paper SPE 12168, presented at the Society of Petroleum Engineers Annual Technical Conference and Exhibition, San Francisco, CA, Oct. 5–8, 1983).

12. D. R. Watkins, L. J. Kalfayan, and S. M. Blaser, "Cyclic Steam Stimulation in a Tight Clay-Rich Reservoir" (paper SPE 16336, presented at the 57th Annual Society of Petroleum Engineers California Regional Meeting, Ventura, CA, April 8–10, 1987).

13. L. J. Kalfayan and D. R. Watkins, "A New Method for Stabilizing Fines and Controlling Dissolution During Sandstone Acidizing" (paper SPE 20076, presented at the Society of Petroleum Engineers 60th Annual California Regional Meeting, Ventura, CA, April 4–6, 1990).

14. F. O. Stanley, S. A. Ali, and J. L. Boles, "Laboratory and Field Evaluation of Organosilane as a Formation Fines Stabilizer" (paper SPE 29530, presented at the Society of Petroleum Engineers Production Operations Symposium, Oklahoma City, OK, April 1995).

15. G. A. Gabriel and S. R. Erbstoesser, "The Design of Buoyant Ball Sealer Treatment" (paper SPE 13085, presented at the Society of Petroleum Engineers Annual Technical Conference and Exhibition, Houston, TX, Sept. 16–19, 1984).

16. D. M. Bilden *et al.*, "New Water-Soluble Perforation Ball Sealers Provide Enhanced Diversion in Well Completions" (paper SPE 49099, prepared for presentation at the Society of Petroleum Engineers Annual Conference and Exhibition, New Orleans, LA, Sept. 27–30, 1998).

17. N. W. Harrison, "Diverting Agents—History and Application," *Journal of Petroleum Technology* (May 1972): 593–98.

18. J. W. Ely, *Stimulation Treatment Handbook* (Tulsa: PennWell Books, 1987).

19. J. P. Gallus and D. S. Pye, "Deformable Diverting Agent for Improved Well Stimulation," *Journal of Petroleum Technology* (April 1969): 497–504.

20. L. R. Houchin *et al.*, "Evaluation of Oil-Soluble Resin as an Acid Diverting Agent" (paper SPE 15574, presented at the Society of Petroleum Engineers Annual Technical Conference and Exhibition, New Orleans, LA, Oct. 5–8, 1986).

21. K. Thompson and R. D. Gdanski, "Laboratory Study Provides Guidelines For Diverting Acid With Foam," *SPE Production & Facilities* (Nov. 1993).

22. N. A. Menzies *et al.*, "Modeling of Gel Diverter Placement in Horizontal Wells" (paper SPE 56742, presented at the Society of Petroleum Engineers Annual Technical Conference and Exhibition, Houston, TX, Oct. 3–6, 1999).

23. G. E. King and G. B. Holman, "Acidizing Quality Control at the Wellsite," booklet (Tulsa: Amoco Production Research Co., 1982).

24. H. O. McLeod and A. W. Coulter, "The Stimulation Treatment Pressure Record—an Overlooked Formation Evaluation Tool," *Journal of Petroleum Technology* (Aug. 1969): 951–60.

25. G. Paccaloni, "New Method Proves Value of Stimulation Planning," *Oil & Gas Journal* (Nov. 19, 1979): 155–60.

26. G. Paccaloni, "Field History Verifies Control, Evaluation," *Oil & Gas Journal* (Nov. 26, 1979): 61–65.

27. L. Prouvost and M. J. Economides, "Real-time Evaluation of Matrix Acidizing Treatments," *Petroleum Science and Engineering* (Nov. 1987).

28. F. R. Behenna, "Interpretation of Matrix Acidizing Treatments Using a Continuously Monitored Skin Factor" (paper SPE 30121, presented at the Society of Petroleum Engineers European Formation Damage Symposium, The Hague, The Netherlands, May 15–16, 1995).

29. A. D. Hill and D. Zhu, "Real-Time Monitoring of Matrix Acidizing Including the Effects of Diverting Agents" (paper SPE 28548, presented at the Society of Petroleum Engineers Annual Technical Conference and Exhibition, New Orleans, LA, Sept. 25–28, 1994).

30. M. J. Economides and K. G. Nolte, editors, *Reservoir Stimulation*, 3d ed. (Schlumberger Educational Services, 2000), chapter 20.

31. C. T. Montgomery, Y.-M. Jan, and B. L. Niemeyer, "Development of a Matrix Acidizing Stimulation Treatment Evaluation and Recording System," *SPE Production & Facilities* 10:6 (Nov. 1995): 219.

32. A. Daneshy, "Economics of Damage Removal and Production Enhancement" (paper SPE 30234, presented at the Society of Petroleum Engineers European Formation Damage Symposium, The Hague, Netherlands, May 15–16, 1995).

7 Unconventional Sandstone And Geothermal Well Acidizing Procedures

One can take a general conventional sandstone acidizing procedure, remove or insert a step, and call it novel or unconventional. Certainly, this has been done many times to sell new products or acidizing systems. However, there have been genuine advancements with unconventional methods introduced to the industry by creative, reasonable risk-taking, stimulation design engineers. Examples of such interesting methods are presented in this chapter. Some examples are:

- Maximum rate/maximum pressure HF acidizing[1–3]
- High-concentration HF acidizing
- CO_2-enhanced HF acidizing[4]
- On-the-fly minimum volume HF acidizing
- Acidizing geothermal (steam) wells[5, 6]

These methods should be considered only for special applications.

MAXIMUM RATE/MAXIMUM PRESSURE HF ACIDIZING

This technique calls for pumping acid treatment stages at as high a rate as possible, but below fracturing pressure. It is postulated that by maintaining maximum injection rate, while always increasing it to maintain the maximum allowable matrix injection pressure, the need for diverter is removed. The technique is known as MAPDIR (maximum pressure differential and injection rates).[1–3]

MAPDIR was a breakthrough in well stimulation when developed by AGIP, the Italian oil concern. It has come under some criticism in recent years because of inherent limitations in application. However, these are unfair in light of the advancement, and considering that many wells have been stimulated successfully with the MAPDIR technique.

The most attractive feature of the MAPDIR technique is that if properly applied, under the proper circumstances, diverting agents are not needed, even in the treatment of long intervals. Successful treatments of intervals as long as 460 ft are reported. Therefore, MAPDIR has been considered a diverting technique, although it really is not a diversion method in the true sense. However, under conditions of limited zone height or length, and limited permeability variation within the zone, the MAPDIR technique has proven more effective than treatment at constant, limited rate, with diversion.

It seems that, in general, for intervals greater than about 150–200 ft, conservatively, sufficient coverage may not be achievable with MAPDIR alone. This is also the case where permeability varies substantially across the treatment zone. However, this is not to say that the MAPDIR technique cannot include a diverter stage, or more than one stage. This can be designed, depending on the known permeability variation and contrasts.

The basic HF acidizing treatment design described in chapter 6 is applicable. Preferred MAPDIR treatments utilize the basic three-stage acidizing design, for example:

1. 15% HCl preflush
2. 12% HCl-3% HF main acid
3. 3% NH_4Cl overflush

Maintaining the maximum rate is critical for success. A maximum differential pressure (*dP*) is maintained throughout the acid treatment. Slightly viscosified acids are often used to lower friction pressure. This allows for higher injection rates with lower horsepower requirements.

The MAPDIR method can be combined with real-time evaluation of skin damage removal. In fact, the calculated skin remaining at any time during treatment is used to determine the maximum allowable differential pressure (*dP*). Therefore, it is used to determine the maximum allowable acid injection rate at any given time, based on the following relationship:

$$\mathrm{dP} = p_{iw} - p_e = 141.2 q_i B\mu / kh \; [(\ln r_e/r_w) + s]$$

where

p_{iw} is the bottomhole injection pressure, psi

p_e is the reservoir pressure, psi

q_I is the production rate, bbl/d

B is the formation volume factor, RB/STB

μ is the viscosity, cp

k is the permeability, darcies

h is the reservoir thickness, ft

r_e is the reservoir radius, ft

r_w is the wellbore radius, ft

s is the skin

Once skin damage is removed, if that can be assessed accurately, pumping should be stopped and the treatment terminated. Skin calculation can at least be performed without confounding it with the presence of diverter.

MAPDIR does require that stimulation fluids be placed downhole to cover the entire interval to be stimulated. This necessitates circulating acid across the interval before pumping fluids away. This may not be possible in certain completions. If it is possible, mechanical isolation of the zone to be treated may be required. Once acid is spotted across the entire interval, the treatment is pumped at the maximum rate and maximum differential pressure below fracturing pressure.

A good use for MAPDIR is in treating long, damaged, naturally fractured intervals. It is recommended for use in longer, open-hole, or liner completions in single or multiple zones, as well as in thin zones. Again, although it is not intended to be used with diverters, the MAPDIR technique could be combined with the use of diverters in such formations. Without diverters, difficulties with the MAPDIR technique exist in tight sands where injection rates may be very limited.

In extremely damaged wells (e.g., wildcat wells), it has been further recommended that if HCl preflush injection cannot be initiated, then the treatment should begin with HF stage injection. (This is done despite the potentially damaging effects of pumping HF without preflush.) The thought is that if the formation is so severely damaged that the HCl preflush cannot be injected, then injection of HF will not make matters worse. It might even serve to establish injectivity through dissolution reactions. If injectivity is established, treatment can continue with HF, or can revert to HCl preflush and proceed as originally planned.

HIGH-CONCENTRATION HF ACIDIZING

There is not a great deal published to support high-concentration HF acidizing, which opposes conventional guidelines. However, there is some evidence that high-concentration HF treatments can be effective in certain sandstone formations.

High-concentration HF means concentration that is high relative to conventional wisdom, not concentrated HF, by any means. For example, high-concentration HF mixtures would be mixtures containing 5%, 6%, or maybe even 9–10% HF, combined with HCl, HCl-organic mixtures, or other proprietary organic-based acids.

The only way to determine the benefit of using a high-concentration HF mixture would be through the use of core flow testing with the actual formation core of interest. So far, the combination of formation and mineralogical properties that lend themselves to high-concentration HF-acidizing cannot be accurately predicted. It is a complex combination, though. Therefore, this is not a technique to be used speculatively.

More often than not, higher HF concentrations will be very damaging to rock competency. This can lead to near-wellbore formation failure, possibly sloughing in perforations, and certainly to sand production.

However, in certain cases, depending on formation characteristics, acidizing with higher concentrations of HF can be successful and not damaging to a significant extent. It would appear that it is the formation of channels, or wormholes, not quite similar to those created in carbonates, but open channels, nonetheless, that are at play with this method. Stimulation with high-HF concentration is probably not often a result of enhanced matrix stimulation within pores and pore throats. Instead, it likely is the result of the creation of channels between grains, as evidenced in core testing.

CO_2-ENHANCED HF ACIDIZING

CO_2-enhanced HF acidizing is a method that involves displacing oil near the wellbore with a gas, preferably CO_2, prior to acidizing.[4] Development of the method by Amoco was based on the observation by Gidley *et al.* that gas wells and oil wells respond differently to the amount of mud acid used in treatment.[7] With gas wells, the stimulation response was found to be in proportion to the amount of mud acid used. However, with oil wells, the response reached a maximum at a specific volume of about 75 gal/ft.

The reason for this difference, as postulated by Gidley, is the greater difficulty in cleaning up oil wells following acidizing. It is believed that crude oil can adsorb on acid reaction products such as the various compounds of silica that reprecipitate on rock surfaces. This can alter wettability from water-wet to partially oil-wet. Also, the fine, reprecipitated silica compounds can form solid-stabilized emulsions or sludges with the crude oil present, either of which can cause plugging.

In CO_2-enhanced HF acidizing, a formation preconditioning stage is employed to displace oil from acid stages. This stage presumably eliminates the formation of emulsions or sludges between spent acid products and the crude oil that would otherwise be contacted. An example treatment procedure for CO_2-enhanced HF acidizing is given in Table 7–1.

Treatment Stage	Treatment Volume
1. Asphaltene stabilizer (solvent)	5–10 gal/ft
2. CO_2 formation preconditioning	100–200 gal/ft
3. Preflush: 15% HCl (injected with CO_2; 50/50)	50 gal/ft
4. Main acid: 12% HCl-3% HF (inj. with CO_2; 50/50)	50–150 gal/ft
5. Overflush: 3% NH_4Cl	10 gal/ft

Table 7-1. *Example Treatment Procedure for CO_2-Enhanced HF Acidizing*

The volumes of preflush and 12-3 (12% HCl-3% HF) mud acid can be varied. Of course, 12-3 mud acid is not required. Other HF mixtures may be used too. In any case, CO_2 is injected with acid. This reduces the density of acidizing fluids. The addition of CO_2 to acid stages also appears to reduce the preferential stimulation of water zones. Acid stages must contain additives for corrosion control and iron control. Amoco also prefers to add clay stabilizer and mutual solvent to the acid stages. Sometimes a surfactant is added as well. Mutual solvent is included in the small volume overflush.

ON-THE-FLY MINIMUM VOLUME HF ACIDIZING

It is sometimes quite a bit to ask of the service companies to expect them to enthusiastically endorse this method, especially offshore or in remote locations. However, it does make sense when employing real-time pressure evaluation methods. This idea depends on calculating skin or estimating it in real-time during the pumping of an acid treatment. When it appears that skin is completely removed, acid pumping stops (i.e., HF stages).

Any acid that was not pumped (because it was not needed) must be taken back. Understandably, the service company does not prefer this. However, in locations where it is not totally impractical, such treatments are viable and agreeable.

The basis of this method is that any acid is wasted that is pumped beyond the point at which skin damage is completely removed. This is because acidizing an undamaged sandstone formation or treatment interval cannot be expected to significantly increase production potential.

Therefore, on-the-fly minimum volume HF acidizing requires that regardless of how much acid is mixed in acid tanks, only that volume absolutely necessary to remove skin should be pumped. This amount is dictated by real-time evaluation. Once skin damage is removed completely, as indicated by the monitoring and evaluation method used, HF acid stage pumping should be terminated. The treatment then should be switched to overflush.

ACIDIZING GEOTHERMAL WELLS

Acidizing geothermal wells is related to sandstone acidizing in that most geothermal reservoirs produce from volcanic rock (andesite). Most formations that produce steam and hot water are naturally fractured. These formations are not water-sensitive, due to existing mineralogy associated with the thermal conditions. Formation temperatures are typically in the range of 400–650 °F. Formation conditions are often conducive to large-volume, high-rate acid treatments.

CANDIDATE SELECTION

Candidate selection should be based on the same principles discussed earlier. In geothermal wells, the strongest indication of acid-removable formation damage is a sharp drop in production rate. In a new well, poor production is the only indication. Well testing techniques are difficult to employ in geothermal wells.

One somewhat informative well testing method, albeit limited, is the fluid entry survey (spinner survey). Location of flow restriction can often be assessed with this test method. Sinker bars of various sizes have also been used to locate wellbore restrictions such as scale, which is common in geothermal producers.

Nearly all geothermal wells that are acidizing candidates have been damaged by:

- Drilling mud solids and drill cuttings lost to the formation fractures
- Scale (calcium carbonate, silica, calcium sulfate, and mixtures)

One thing geothermal wells have in their favor is that complete damage removal is not necessary. Partial removal of damage with acid treatment may eventually result in complete damage removal when the treated well produces back. The high-rate and high-energy backflow from geothermal wells can blow out damage that was not dissolved by acid. Damage that was softened, broken up, or detached from downhole tubulars and fracture channels can be produced back through a large diameter casing completion.

Erosion of production lines may occur if drill cuttings are produced back during blow down of a well after stimulation. Care must be taken in this regard. A temporary flow line may be required until solids production has stopped.

ACID TREATMENT DESIGN

A very successful method of acidizing geothermal wells has been a basic, high-rate, brute-force method. High acid concentrations have been shown to be effective in geothermal wells producing from natural fractures not containing separate, large carbonate zones. An example of a basic treatment procedure for a geothermal well completed open hole or with a liner is given in Table 7–2.

Treatment Stage	Treatment Volume
1. Preflush and cooling	Fresh water: 10 bpm (through tubing and annulus—cool wellbore temperature to below 200 °F)
2. Acid preflush: 15% HCl	10,000–50,000 gal; 2–10 bpm
3. Main acid: 10% HCl-5% HF	10,000–50,000 gal; 2–10 bpm
4. Overflush: fresh water or 3% NH_4Cl	Several hours at maximum rate

Table 7-2. *Example of a Basic Treatment Procedure for a Geothermal Well (Completed Open Hole or with a Liner)*

In geothermal well acidizing, more acid often is better. Naturally fractured volcanic formations can withstand high HF concentration. The HCl-HF stage can be 10% HCl-5% HF, or 10% HCl-7% HF, for example. These acid mixtures have been used successfully in stimulating geothermal wells in southeast Asia (the Philippines), where a large number of acid treatments have taken place. Acid volumes can vary quite a bit.

These acidizing treatments have also employed an acid formulation containing 3% HCl-5% HF and an organophosphonic acid.[7] The mixture is less corrosive and may help slow scale reprecipitation, as the phosphonic acid complexes with certain cations in spent acid.

An example treatment design using such a formulation is given in

Table 7–3.

Treatment Stage	Treatment Volume
1. Preflush and cooling	Fresh water: High rate to cool wellbore below 200 °F.
2. Preflush: 10% HCl	24,000 gallons
3. Main acid: 3% HCl-5% HF + 1.5% organophosphonic acid	40,000 gallons
4. Overflush: 3% HCl or 3% NH_4Cl	Wellbore displacement volume
5. Overflush: Fresh water (field water)	Several hours at maximum rate

Table 7-3. *Example of an Organophosphonic-HF Acid Treatment Design for a Geothermal Well*

Acid additives

The only acid additives necessary in a geothermal acid job are:

1. Corrosion inhibitor and inhibitor intensifier (often required)
2. High-temperature iron-control (reducing) agent

Water-wetting surfactants, necessary in oil well stimulation, are not needed in geothermal wells because of the absence of hydrocarbons. Suspending agents (nonemulsifier surfactants) are also not needed, although they seem to be included often in geothermal well stimulation job proposals. Clay stabilizer is not needed.

ACID PLACEMENT

Conventional acid placement techniques are ineffective for the long, open-hole or liner-completed intervals typically encountered in geothermal wells. High-temperature foam systems may improve zone coverage. Gelling agents for thickening acid have been shown to be ineffective in geothermal liner completions. The best way to maximize acid coverage in geothermal wells is by pumping at maximum injection rates.

SAFETY REQUIREMENTS

Certain additional and special precautions must be taken in a geothermal acid job, especially in the remote or jungle locations common to geothermal fields. The safety of local inhabitants who reside near the well sites is of special concern. All empty chemical containers must be cleaned and disposed of immediately.

Care must be taken in minimizing odor problems during the mixing of stimulation fluids. Odor of return fluids following an acid job is occasionally a problem. The large overflush volumes used in geothermal acidizing help in minimizing odor. However, other means are important to make sure that an odor problem is controlled. These can include careful well blow down and monitoring odor at various locations within a certain radius of the well and production system.

REFERENCES

1. G. Paccaloni and M. Tambini, "Advances in Matrix Stimulation Technology," *Journal of Petroleum Technology* (March 1993): 256.

2. G. Paccaloni, M. Tambini, and M. Galoppini, "Key Factors for Enhanced Results of Matrix Stimulation Treatments" (paper SPE 17154, presented at the Society of Petroleum Engineers Formation Damage Control Symposium, Bakersfield, CA, Feb. 8–9, 1988).

3. G. Paccaloni, "A New, Effective Matrix Stimulation Diversion Technique," *SPE Production & Facilities* (August 1995): 151–56.

4. J. Gidley, E. Brezovec, and G. King, "An Improved Method for Acidizing Oil Wells in Sandstone Formations" (paper SPE 26580 presented at the Society of Petroleum Engineers 68th Annual Technical Conference, Houston, TX, Oct. 1993).

5. P. H. Messer, D. S. Pye, and J. P. Gallus, "Injectivity Restoration of a Hot-Brine Geothermal Injection Well," *Journal of Petroleum Technology* (Sept. 1978): 1225.

6. G. di Lullo and P. Rae, "A New Acid For True Stimulation of Sandstone Reservoirs" (paper SPE 37015, presented at the Society of Petroleum Engineers Asia Pacific Oil and Gas Conference, Adelaide, Australia, Oct. 28–31, 1996).

7. J. L. Gidley, "Acidizing Sandstone Formations: A Detailed Examination of Recent Experience" (paper SPE 14164, presented at the Society of Petroleum Engineers 60th Annual Technical Conference and Exhibition, Las Vegas, NV, Sept. 1985).

8 Sandstone Acidizing In Horizontal Wells

Sandstone acidizing in horizontal wells is difficult because of placement limitations. Horizontal zones are typically longer than zones in vertical wells. Horizontal intervals can be many thousands of feet. Mechanical isolation with packers, which is the most effective method of placing acid in vertical wells, is not often practical in horizontal zones. Too many packer settings are required, extending beyond the limits that current commercial packer systems can be effectively seated and reseated.

Diverter choices are also limited. Ball sealers are sometimes not effective in horizontal wells, because they tend to the bottom side of the pipe. Gel and foam diversion can be effective, at least to some extent. In any case, it is known that some attempt to divert acid fluids with a nondamaging diverter is better than no diversion at all.

Economides et al. presented a clever method for diverting acid in horizontal completions.[1] The method uses coiled tubing to pump the acid treatment. Nonreactive fluid is pumped through the annulus between the coiled tubing and production tubing to create back pressure. This dual action forces the acidizing treatment fluids into the formation at or below the end of the coiled tubing string.

This technique has also given rise to the idea of pumping viscous diverter stages down the annulus, while acid is pumped through coiled tubing, in both sandstones and carbonates. During treatment in this manner, coiled tubing can be reciprocated, or drawn along the treatment interval at rates corresponding to desired treatment volumes to be pumped per foot. The acid injection volumes can be varied according to severity of damage, if known, or estimated in certain portions of the horizontal interval.[2]

The MAPDIR technique can be extended to use in horizontal well acidizing. The MAPDIR technique described previously in chapter 7 also has been successfully applied, with modifications, to horizontal wells.[3] An optimized treatment procedure consists of placing acid downhole through coiled tubing. The acid treatment is bullheaded at maximum bottomhole differential pressure and injection rate. About 30% of the total acid volume is first placed across the interval with coiled tubing. The remaining acid is injected at maximum downhole differential pressure and maximum rate.

As in the use of the MAPDIR technique in vertical wells, equations have been developed that link measured injection pressure and rate to evolving skin.[3]

Diversion of fluids can be achieved by sustaining maximum differential injection pressure with maximum acid injection rates. Again, slightly viscosified fluids are used to reduce friction.

Placement of acid across the entire zone with coiled tubing is recommended with the MAPDIR technique. Coiled tubing is placed in the horizontal section and reciprocated while acid is spotted across the zone. This is a somewhat controversial step. Some operators, for good reason, will not place coiled tubing in the horizontal or highly deviated section of a well. The risk of getting stuck may be too high.

Successful high-rate/high-pressure treatments in the offshore California fields have also been reported.[4] It is encouraging that in these treatments, acid was not initially placed across the interval with coiled tubing. The entire acid stage volumes were simply bullheaded from the outset. This was done to avoid the risk of running coiled tubing into highly deviated or horizontal well sections.

These treatments were conducted in naturally fractured Monterey shale. Based on DST results, formation damage was suspected. Evaluation of previously unsuccessful acid jobs in the same zones, drilling records, workover data, and laboratory testing suggested that formation damage was due to one or more of the following:

- Mud invasion
- Solids ($CaCO_3$ scale) precipitation
- Emulsion blocks in the formation

Not until the high-rate/high-pressure method was used was successful stimulation accomplished. Resulting production increases have been from 85–100%. The general treatment design is given in Table 8–1.

Stage	Treatment
1. Surfactant soak	To restore formation to water-wet state (1–2 days)
2. Acid preflush	HCl/acetic mixture
3. Main acid	12% HCl-3% HF
4. Overflush	3% NH_4Cl

Table 8–1. *High-Rate, High-Pressure Treatment Design*

Reported volumes of acid used are more than 100 gal/ft of formation height. A typical minimum pump rate is 7 bpm. A rate of up to 8.5 bpm has been required to maintain maximum differential pressure.

REFERENCES

1. M. J. Economides, K. Ben Naceur, and R. C. Klem, "Matrix Stimulation Method for Horizontal Wells," *Journal of Petroleum Technology* (July 1991): 854–61.

2. M. J. Economides and T. P. Frick, "Optimization of Horizontal Well Matrix Stimulation Treatments" (paper SPE 22334, presented at the Society of Petroleum Engineers International Meeting on Petroleum Engineering, Beijing, China, March 24–27, 1992).

3. M. Tambini, "An Effective Matrix Stimulation Technique For Horizontal Wells" (paper SPE 24993, presented at the Society of Petroleum Engineers European Petroleum Conference, Cannes, France, Nov. 1992).

4. M. Juprasert, "Bullhead Acidizing Succeeds Offshore California," *Oil & Gas Journal* (April 11, 1994): 47.

part three

carbonate acidizing

9 Purposes of Carbonate Acidizing

There are two basic types of acidizing treatments applicable in carbonates. They are characterized by injection rates and pressures. Treatments with injection rates below formation fracturing pressure are called matrix acidizing treatments. Matrix acidizing is applicable only in formations exhibiting formation damage. There are exceptional cases in which acidizing at matrix rates in an undamaged carbonate formation may result in an acceptable stimulation response. Naturally fractured carbonate formations, in particular, are examples.

Treatments injected at rates above fracturing pressure are termed fracture acidizing, or acid fracturing treatments. Fracture acidizing is applicable in both damaged and undamaged carbonate formations. Fracture acidizing is a treatment in which fracturing is initiated and propagated by a suitable fracturing fluid. The walls of the fracture are then etched with acid to create a conductive flow channel upon formation closure.

Acid fracturing describes a treatment in which the fracturing and etching fluids are acids. The fracture is created with a viscous acid system, for example, which also etches the walls of the fracture during the injection process.

These days either term (fracture acidizing or acid fracturing) is used to describe the process of creating an acid-etched fracture in a carbonate formation.

MATRIX ACIDIZING

The primary purpose of matrix acidizing is to improve flow capacity through a damaged region near the wellbore. This can be achieved by dissolving rock through near-wellbore formation damage.

Most matrix acidizing treatments in carbonates use hydrochloric acid mixtures of one kind or another. Hydrochloric acid dissolves limestone and dolomite to produce open, conductive channels. If extended, these channels form wormholes, which can bypass near-wellbore formation damage. The effective treated zone can become much larger than in sandstones.

Retarded acid systems can extend the length and number of wormholes. Such systems include slightly gelled acid, chemically modified acid, surfactant-retarded acid, emulsified acid, and foamed acid. However, the time it takes for acid to spend is still short in most cases. More often than not, only the formation near the wellbore can be treated effectively. Thus, effective uniform matrix treatment beyond several feet from the wellbore is unusual.

Acid-removable skin, or in the case of carbonates, skin that can be bypassed by acid, must be present for a matrix acidizing treatment to be effective, realistically.

FRACTURE ACIDIZING

Fracture acidizing treatments of carbonates are conducted for the purpose of either bypassing formation damage or stimulating undamaged formations.

As mentioned in chapter 2, fracture acidizing is an alternative to hydraulic propped fracturing in carbonates. With fracture acidizing, fracture conductivity is achieved by acid-etching the walls of the created fracture. With hydraulic propped fracturing, fracture conductivity is achieved by filling the fracture with solid proppant to hold it open.

There are two general methods of achieving an acid-etched fracture: (1) viscous fingering (pad-acid) and (2) viscous acid fracturing. With the viscous fingering, or pad-acid method, a fracture is first created with a viscous gelled water pad. Acid with lower viscosity is then injected into the created fracture. The lower viscosity acid "fingers" through the viscous pad rapidly and unevenly, thereby penetrating deeply and etching the fracture face unevenly.

The viscous fingering method is applicable in all carbonates but is most useful in more homogeneous formations, such as higher purity limestones and chalks. It was developed for such formations in which uneven or differential acid-etching may not occur unless induced.

Viscous acid fracturing uses viscous acid systems such as gelled, emulsified, and foamed acid, or chemically retarded acids, to both create the fracture and differentially etch the fracture face. Treatments with viscous acid are applicable in heterogeneous carbonates such as dolomites or impure limestones.

Purposes of carbonate acidizing include:

- Perforating fluid, perforation cleanup, and break down
- Damage bypass (matrix acidizing, fracture acidizing)
- Formation stimulation (fracture acidizing)

PERFORATING FLUID, PERFORATION CLEANUP, AND BREAKDOWN

In carbonate formations, acetic acid is an effective perforating fluid (usually 9% or 10%). It may also contain salt to weight up, if required. Acetic is a mild acid, but strong enough to dissolve perforation debris, resulting in clean perforations. If additionally necessary following well completion, perforations may be cleaned or stimulated further with a small acid treatment, typically acetic (10–15%) or HCl (7.5–15%).

Perforation breakdown, conducted to initiate production, may also be performed. Acid is injected above fracturing pressure to break down the perforations and create communication between the formation and the wellbore.

If a well is to be completed with a hydraulic fracturing treatment, perforation or formation breakdown prior to injection of fracturing fluids is a common need and practice, as well. Injection can be initiated by pumping thin acid (usually HCl) or a "slick" HCl solution (slightly gelled) to break down perforations and initiate fluid entry in the formation. This enables easier placement of fracturing fluids and proppant in a hydraulic fracturing treatment.

Damage bypass. Damage bypass in carbonate formations can be achieved with either of two methods applicable to carbonate acidizing:

- Matrix acidizing
- Fracture acidizing

Matrix acidizing will form wormholes through, and hopefully beyond, the damage radius, thereby bypassing formation damage. Wormholing can be influenced by the nature of the acid system used. There are slower-reacting acids, such as acetic or EDTA, as well as retarded acid systems, such as chemically (surfactant) retarded HCl. These acid systems will tend to form relatively shallower, but more branched wormholes under similar conditions (temperature, injection rate, etc.).

Strong acid, such as HCl, will typically form longer, single wormholes extending from the perforations. If acid is not effectively diverted, a matrix treatment with conventional HCl may create only one wormhole. This wormhole will then accept all acid injected, as it will be the increasingly conductive path of least resistance for acid as it continues to be injected.

Fracture acidizing also can be used to bypass formation damage in a carbonate formation. For damage bypass, a long fracture may not be needed. The purpose would be similar to that of a frac-and-pack procedure in a sandstone formation. In that case, a propped fracture with length and conductivity sufficient to effectively extend the wellbore radius beyond a damage zone is all that is needed—not an extensive, propped fracture.

FORMATION STIMULATION

In a carbonate formation, stimulation of the formation itself realistically can only be achieved with fracture stimulation methods. These fracture stimulation methods can be either hydraulic fracturing or fracture acidizing (acid fracturing). To stimulate an undamaged formation, an extensive conductive flow path deep into the formation must be created. This can only be accomplished through fracturing.

Two key factors in the success of an acid fracturing treatment are the resulting etched fracture length and conductivity. The effective fracture length is influenced by: (1) acid fluid-loss characteristics; (2) acid reaction rate; and (3) acid flow rate in the fracture. Acid fluid loss is the most important factor affecting fracture length.

Conductivity is largely dependent on how the fracture faces are etched. The fracture faces must be etched in a nonuniform manner to create conductive flow channels that remain open after fracture closure. Generally, good conductivity results from formation heterogeneity and flow-induced selective etching.

Although fracture length and conductivity are difficult to predict, acid fracturing, like hydraulic fracturing, can substantially improve productivity and, in certain cases, increase field oil and gas reserves.

10 Some Comments On The Chemistry And Physics Of Carbonate Acidizing

The chemistry and physics of carbonate acidizing are exhaustive topics. On one hand, the chemistry is simple in terms of reactions and reaction products. On the other hand, the chemistry is complex in terms of the reaction kinetics of various acids and acid mixtures that can be used. Carbonate acidizing physics and mathematics are fascinating but very involved. They can be complex, uncertain, and depressing.

Fortunately, both the chemistry and physics of carbonate acidizing have been studied, reported on, and comprehensively reviewed by many others. The majority of this great body of work is theoretical, or confined to the laboratory. Much is controversial.

Most of the work in this area is outstanding and has led to the development of our current understanding of the remarkable carbonate acidizing processes. However, these studies and their results can and almost certainly do fall a bit short of reality—what actually happens in the formation. We really cannot know exactly what happens in the formation. Actual field results, and laboratory results too, for that matter, consistently indicate that it is still not possible to predict and design carbonate acidizing treatments definitively. But that is the nature of it, and it should not be criticized. We take what we can use and apply it creatively and intelligently.

There are excellent references on these topics, several of which are cited herein. The reader is encouraged to explore them and other related material.

CHEMISTRY

The chemistry of carbonate acidizing is fundamentally different than that of sandstone acidizing.[1–3] For one thing, pure carbonate minerals react rapidly and completely with excess HCl, yielding water, CO_2, and highly soluble chloride salts. Siliceous minerals are typically only slightly soluble in hydrochloric acid. Of course, hydrofluoric is the operative acid in sandstone acidizing.[4–6] HF acid is required to dissolve plugging by siliceous particles, and its reaction with certain minerals, such as feldspars, is catalyzed by HCl.[7]

HF acidizing finds no application in carbonates, as it forms solid calcium fluoride (CaF_2) in limestone, and both calcium fluoride and magnesium fluoride (MgF_2) in dolomite. In any case, HF reaction in sandstones cannot be considered analogous to HCl reaction in carbonates. HF reaction in sandstones is controlled by the surface area of siliceous minerals—the surface reaction kinetics.

HCl reaction in carbonates is controlled by the mass transport of acid to the mineral surfaces. In sandstones, acid transport rate is high compared to surface reaction rates, and in carbonates, surface reaction rates are high compared to acid transport rate. The slower rate step (acid transport or surface reaction) will control the reaction kinetics. Overall, HCl reactions in carbonates are much faster than HF reactions in sandstones.

Another difference between acidizing in carbonates and sandstones is that the chemistry of carbonate acidizing is generally straightforward. In contrast, the chemistry of sandstone acidizing is quite complex, both with respect to dissolution (stimulation) reactions and reprecipitation (damaging) reactions.

In the acidizing of carbonates, both matrix and fracture, dissolution reactions are simple, and acid/formation reprecipitation reactions are not of much concern, as they most certainly are in sandstone acidizing. Reprecipitation of acid reaction products in carbonates can occur with certain organic acids, such as citric and formic, but only if they are used in excess concentration. There is no reasonable application of citric acid in carbonate acidizing where this would occur.

Use of formic acid in concentrations greater than about 10% can precipitate the reaction product calcium formate. Also, use of HCl in concentrations greater than 20% can sometimes form certain insoluble calcium compounds.

In matrix acidizing of carbonates, HCl is so rapidly and completely reactive with carbonate minerals that macroscopic channels are formed through the rock matrix. The formation of wormhole channels is also accomplished with acetic and formic acids, which are weaker organic acids. Wormhole channels can also be created with calcium-chelating agents, such as EDTA, and acid/acid precursors, such as chlorinated acetic acids.

Years ago, even certain chlorinated hydrocarbons (e.g., carbon tetrachloride and allyl chloride) were considered for higher temperature wells. Such compounds would hydrolyze downhole to generate HCl. This was a clever concept for generating acid in situ. However, chlorinated hydrocarbons can no longer be considered because many are classified as carcinogens.

In fracture acidizing, acid-etching of the fracture requires sufficient acid strength and dissolving power. Realistically, these features are provided only by HCl or, in certain cases, organic acids. Chelating agents do not dissolve sufficient volumes of rock to be applicable to fracture acidizing.

Acids Used in Carbonate Acidizing

The common acids used to stimulate carbonate formations are:

- Hydrochloric (HCl)
- Acetic (CH_3COOH)
- Formic (HCOOH)

Hydrochloric acid. The most common field solution for matrix and fracture acidizing application is 15% HCl. Concentrations may vary between 3% (for tubing cleaning application) and 35%, but they typically are between 15% and 28%. Acid solutions containing greater than 15% HCl are called high-strength acids.

Typically, high-strength acidizing treatments use 20% HCl or 28% HCl.

Acetic acid. Acetic acid is a weakly ionized, slow-reacting organic acid. Acetic acid is easier to inhibit against corrosion than HCl. For this reason, acetic acid has application as a perforating fluid in carbonate wells and as a stimulation fluid in high-temperature formations.

Acetic acid has other advantages over HCl. It naturally sequesters iron, reducing iron precipitation. It is useful in stimulating wells with alloy tubulars. Acetic acid does not attack chrome plating below 200°F.

Typical acetic acid concentration is 10%.

Formic acid. Formic acid, like acetic, is a weakly ionized, slower-reacting organic acid. It is a stronger acid than acetic and somewhat more difficult to inhibit against corrosion. However, formic acid corrosion of steel is of the nonpitting variety; therefore, it can be used in higher temperature applications, as well. It is especially useful when mixed with acetic acid. Acetic/formic blends can be effective stimulation fluids for deeper, higher temperature carbonates.

Typical formic acid concentration is 9% or 10%.

REACTIONS OF ACID WITH LIMESTONE AND DOLOMITE

The factors that affect the spending rate of acid in carbonate formations are:[8,9]

- Temperature
- Pressure
- Acid type
- Acid concentration
- Acid velocity
- Reaction products
- Area-volume ratio
- Formation composition (structure and mineralogy)

Temperature. Acid reaction rate increases with temperature. At about 150°F, the reaction rate of HCl and limestone is about twice that at 80°F.

The reaction rate of HCl in limestone is faster than in dolomite up to about 250°F (or somewhat less). At higher temperatures, the rates of reaction in limestone and dolomite are equally fast.

Pressure. Pressure greater than 500 psi (pounds per square inch) has little effect on reaction rates. Below 500 psi, increased pressure increases reaction rate.

Acid type. Acid strength varies with acid type. Acid strength is defined by ionization strength, or the degree to which acid ionizes to hydrogen ion (H^+). The hydrogen ion is the reactive species with carbonate minerals, not the acid molecule. The acid ionization reactions are:

$$HCl + H_2O \rightarrow H^+ + Cl^-$$
$$H_3C\text{-}COOH + H_2O \rightarrow H^+ + H_3C\text{-}COO^-$$
$$HCOOH + H_2O \rightarrow H^+ + HCOO^-$$

HCl is the strongest of these acids, as it is nearly completely dissociated to H^+ and Cl^- in water. Acetic and formic acids are weakly ionizing, as they do not completely dissociate to H^+ and the corresponding anion in water. Consequently, organic acids have a lower spending rate, and because of their higher equivalent weights, will have less dissolving power for the same percentage acid solutions.

Acid concentration. HCl acid spending rate reaches a maximum at a concentration of about 20%. At higher concentrations, the large volumes of $CaCl_2$ and CO_2 generated in solution have the effect of retarding reaction. Acetic and formic acids are naturally retarded by their reaction products, calcium acetate and calcium formate, respectively. The benefit from increased concentration diminishes.

Acid velocity. Velocity has some effect on spending rate. In acid fracturing, increasing velocity increases live acid penetration.

Area/volume ratio. The spending rate of acid is proportional to the surface area of rock that comes in contact with a given volume of acid. In matrix acidizing, the surface area/acid volume ratio is very high, and acid spends rapidly. It is therefore very difficult to achieve deep penetration of live acid in matrix stimulation treatments unless the acid is extremely retarded, which can compromise treatment effectiveness.

In natural fractures, the surface area/acid volume ratio is much less, and deeper treatment is possible. In acid fracturing, the ratio of rock surface area/acid volume is even lower. Very deep stimulation in fracturing applications is therefore possible, especially with slower reacting, low leakoff acid systems.

Formation composition. The chemical and physical compositions of the formation are very important in defining acid spending time, and, subsequently, acid penetration distance. Acid spends very rapidly in highly reactive carbonates (>95%). Acid spending time can be much slower in formations with lower HCl reactivity (65–85%). As mentioned previously, the reaction rate of acid in limestone is about twice that in dolomites (at lower temperatures). Therefore, live acid penetration can be deep in low solubility, lower temperature dolomites.

Acid reactions with carbonates are represented stoichiometrically as follows:

1. Hydrochloric acid + calcium carbonate → calcium chloride + carbon dioxide + water

$$2HCl + CaCO_3 \rightarrow CaCl_2 + CO_2 + H_2O$$

2. Hydrochloric acid + dolomite → calcium chloride + magnesium chloride + carbon dioxide +water

$$4HCl + CaMg(CO_3)_2 \rightarrow CaCl_2 + MgCl_2 + 2CO_2 + 2H_2O$$

3. Acetic acid + calcium carbonate → calcium acetate + carbon dioxide + water

$$2HCH_3CO_2 + CaCO_3 \bullet Ca(CH_3CO_2)_2 + CO_2 + H_2O$$

4. Acetic acid + dolomite → calcium acetate + magnesium acetate + carbon dioxide + water

$$4HCH_3CO_2 + CaMg(CO_3)_2 \rightarrow Ca(CH_3CO_2)_2 + Mg(CH_3CO_2)_2 + 2CO_2 + H_2O$$

5. Formic acid + calcium carbonate → calcium formate + carbon dioxide + water

$$2HCOOH + CaCO_3 \rightarrow Ca(HCO_2)_2 + CO_2 + H_2O$$

6. Formic acid + dolomite → calcium formate + magnesium formate+ carbon dioxide + water

$$4HCH_3CO_2 + CaMg(CO_3)_2 \rightarrow Ca(HCO_2)_2 + Mg(HCO_2)_2 + 2CO_2 + 2H_2O$$

With the weaker acids, acetic and formic, the dissociation of the hydrogen ion is also suppressed by the generation of CO_2, a weak acid in solution. In HCl-organic acid blends, especially at high temperature, the organic acid may therefore not contribute much, if any, formation dissolution. This is true until the acid is considerably spent (HCl has reacted to near-completion); at that point, the organic acid will dissociate.

Table 10–1 shows relative dissolving power in terms of pounds of formation (limestone or dolomite) that common mixtures of these acids will dissolve.

Formic acid at 10% concentration has the same dissolving power as 8% HCl. A 10% acetic acid solution has the same dissolving power as 6% HCl. Formic acid concentration must be limited to about 10%. Above that level, calcium formate precipitates. Acetic acid can be used at much higher concentrations, as calcium acetate remains soluble. However, the common mixture is 10% acetic. This is practical given the relative reactivity and the fact that the organic acids are more expensive than HCl.

None of these acids react appreciably with sand (SiO_2). They will react with iron compounds or minerals containing iron. Acetic acid forms a complex with iron in solution and helps prevent iron precipitation. Iron content

Acid	Zero porosity limestone	Zero porosity dolomite
15% HCl	1.84 lbs/gal	1.71 lbs/gal
10% Formic	0.92 lbs/gal	0.85 lbs/gal
10% Acetic	0.72 lbs/gal	0.66 lbs/gal

Table 10–1. *Carbonate Acidizing Dissolving Power*

in the formation, and its source, must be considered in carbonate acid treatment design. In formations with high iron content, it may be necessary to cut back on HCl concentration or substitute part or all HCl with acetic acid.

Chelating agents, such as EDTA, are even weaker acids than acetic and formic acids. A 9% disodium EDTA solution, for example, has the dissolving power of 2.2% HCl, approximately. As mentioned before, the chelating agents do not have application in fracture acidizing. However, their use in matrix acidizing is intriguing because they can impart favorable wormhole or channeling structures to carbonates.[10] They are not often used, presumably because of their cost relative to the common acids (HCl, acetic, and formic) and lower dissolving power. The lower dissolving power is not necessarily a disadvantage.

PHYSICS

The physics of acid fracturing and matrix acidizing in carbonates are both complex.

ACID FRACTURING

Acid fracturing response depends on final fracture length and conductivity. Length depends on live acid penetration distance, and conductivity depends on the etching pattern imparted on the fracture walls by acid.

Live acid penetration distance is enhanced by reducing the mass transfer or diffusion of acid in the fracture to the reactive fracture wall surfaces. This slows or partially blocks the acid reaction itself, reducing fluid loss (leak-off) from the fracture to the matrix. Fluid loss reduction has the greatest effect. Fluid loss is controlled by several factors, including formation permeability and porosity, leak-off fluid viscosity, compressibility of the reservoir fluids, and the difference in pressure between the fracture and the matrix.

Leak-off is more severe in formations with high permeability and in gas wells (high reservoir fluid capacity). Leak-off occurs through acid-created channels, or wormholes, branching off from the fracture to the matrix. Leak-off can be partially controlled by reducing the wormhole creation (through acid retardation) or physically blocking wormholes created.

The etching pattern, or effective etched width, is affected by:

- Mass transport of acid from the body of the fracture to the walls of the fracture
- Reaction of acid on the rock surface
- Leak-off of acid from the fracture to the formation matrix
- Heat transfer in the fracture

Viscosifying acid for retardation can favorably influence one or more of these factors. Viscous acids include gelled, emulsified, foamed, and surfactant-thickened acids. The intent of viscosifying acid is to slow the rate of acid diffusion to the rock surfaces and to reduce fluid loss. Work by Crowe, *et al.* indicated, however, that gelling acid does not have a significant effect on reducing mass transfer of acid.[11]

Additional fracture width from the higher viscosity may play a role in increasing acid penetration, as well as reducing leak-off, which is still expected to be high. Gel polymer can be crosslinked for increased viscosity and to further reduce leak-off to some degree.

The system that may provide the deepest penetration is an oil-external emulsified acid. Acid-oil emulsions can be oil-external or acid-external. The oil-external emulsions have higher dissolving capacity per unit volume and are generally more effective. A common emulsion mix is 70% acid and 30% oil. A limitation is that emulsified acid can be difficult to pump at sufficient rate in deeper wells because of high friction pressure during injection. Gelled acids provide the most friction reduction. The temperature limit of emulsified acid is about 270–300°F.

Foamed acid can be useful in increasing effective fracture length, as well as improving contact in longer treatment intervals. Foamed acid is essentially gas-in-acid emulsion stabilized with a foaming agent. The amount of gas in the foam on a volume basis is called the quality; a foam comprised of 70% gas and 30% liquid is a 70-quality foam. The gas phase is usually nitrogen, but CO_2 may also be used. Most foamed acids are 60–75 quality.

The downside to foamed acid is the reduced amount of acid available per unit volume of fluid. In practice, this is often partially compensated for by using higher strength acid, such as 28% HCl.

Surfactant-thickened acids are systems in which special surfactants are used that viscosify live acid mixtures, which then thin when acid spends.[12] The thickened live acid can enhance penetration, and the thinning with spending improves flowback after treatment. However, the thinning also can increase leak-off, which reduces etched fracture length.

Fluid loss also can be reduced somewhat by physically blocking acid-created leak-off channels. A gelled pad fluid (water gelled with guar-based polymer) similar to conventional hydraulic fracturing fluid, alternated with acid, is the method described by Coulter.[13] Pad stages enter and temporarily block the acid-created wormholes branching off into the matrix. The sealing of these leak-off channels allows live acid to proceed further down the fracture.

ACID FRACTURE DESIGN SIMULATION

Commercial software programs that model the fracture acidizing processes exist. They use the same fundamental prediction methods used in hydraulic fracturing simulation with nonreactive fluids. Modifications for acid reaction kinetics, etc., are included in acid fracturing modules. Settari provides an example and discussion.[14]

Unfortunately, as complex and rigorous as the modeling and prediction of acid fracture geometry and final conductivity can be, all is meaningless if the greatest unknown, leak-off, is not accurately estimated. And it rarely can be. Defining leak-off properties of reactive acid, which is changing in viscosity, density, and temperature in a fracture and formation of unknown leak-off characteristics itself, is not realistic.

Due to the difficulty of accurately estimating leak-off, service company simulation results are often very overly optimistic. Unfortunately, service company engineers are often forced by operators to provide justification for treatment design in a competitive bidding situation. This expectation further encourages generation of fictional fracturing simulation runs.

What is important to know here is that successful fracture acidizing stimulation treatments are common. The most important factor is to incorporate methods using or combining fluid systems that reduce fluid loss, or leak-off. Fracturing with viscous acid systems, or alternating stages of viscous pad and viscous acid, are both practical. Specific design and fluid choices can be arbitrary. This is fine, too, as options are open.

MATRIX ACIDIZING

The physics of carbonate matrix acidizing is also complex. The operative mechanism in matrix acidizing of carbonates is the creation of conductive channels, or wormholes, extending from perforations or the formation face into the formation matrix. Wormholing phenomena have been studied, described, and modeled extensively.[8, 9, 15–31] Given the complex nature of wormhole formation, there is, understandably, some controversy in the modeling of wormholes. Fredd and Miller provide a comparison of various models.[32]

There is a wide range of possible channeling patterns or structures that can be created as acid reacts in carbonates under matrix conditions. The channeling structure depends on injection rate and acid reactivity with the rock, which is a function of the stimulation fluid properties and temperature. For each formation and its conditions, there will exist an optimum combination of acid (or reactive fluid) injection rate and degree of reactivity (or retardation).

Fredd categorizes and describes the major wormhole structure possibilities as:[32]

- Face dissolution (no wormholing)
- Conical wormholes (single channel with limited branching)
- Dominant wormholes (primary channels with some branching)
- Ramified wormholes (extensive branching)
- Uniform dissolution

Figure 10–1 gives a set of rough depictions of these different dissolution patterns and wormhole structures, imagined to be extending from a wellbore (or perforation) at the left of the diagram.

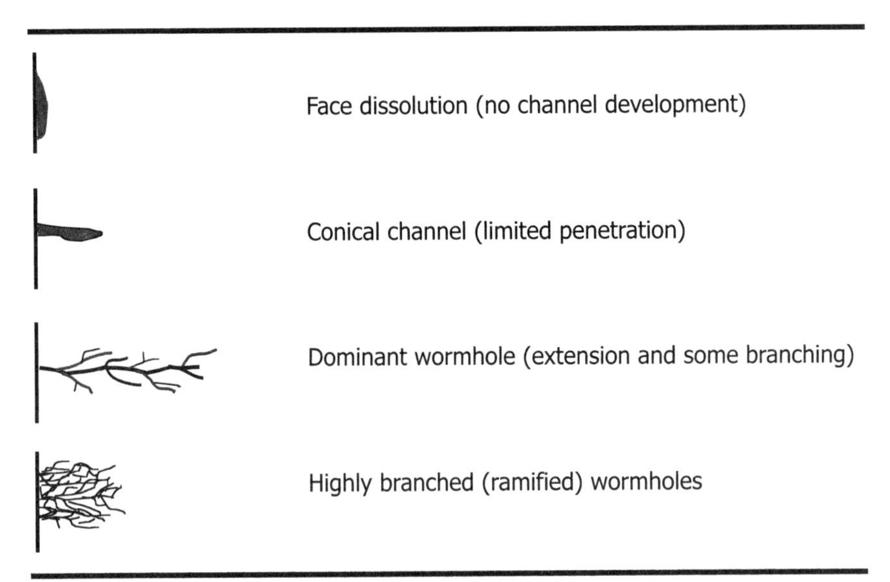

Fig. 10–1. *Dissolution patterns and wormhole structures*

A recent summary by Fredd is comprehensive and includes more detailed pictorial representations of various dissolution patterns and wormhole structures, including those observed in laboratory core flow tests.[33]

Face dissolution

At low injection rates and high reactivity, acid does not penetrate formation adequately and simply dissolves away formation at the face. There is no significant channeling away from the wellbore. This removal mechanism is very inefficient and cannot be expected to stimulate beyond damage.

Conical channel

The formation of a conical channel occurs with a somewhat higher injection rate and/or somewhat lower reactivity. Acid reacts primarily on the walls of the initial channel formed, resulting in an essentially single, wide channel with conical shape. A conical channel formation may lead to a single lengthy wormhole with continued acid volume injected. This pattern is typical of dissolution by strong acid (e.g., HCl) that is not retarded, even injected at high rate. This is due to the HCl being consumed on the walls of the channel as it continues to flow, rather than entering other pore channels or extending to the tip substantially.

Dominant wormhole

At higher injection rates, and with retarded or slower-reacting acids, the stimulation fluid has the opportunity to further extend the tip of the channel being formed. This gives rise to a thinner, longer wormhole. There will be some branching, as reactive fluid also can begin entering smaller pores at the higher flow rate.

Generally, the dominant wormhole pattern represents the optimum combination of injection rate and reactivity. This wormhole pattern has the best chance of extending beyond the damage zone. It has sufficient conductivity and branching to provide significant production flow to the wormhole and into the wellbore.

Highly branched (ramified) wormholes

At very high injection rates, the structure can become much more highly branched, or ramified. Branching can become very extensive, even dominant, as acid can enter smaller pores, thereby branching off the original path significantly. Like the conical channel, or single wormhole pattern, this too is a less-efficient dissolution pattern, as penetration through the damage zone can become less likely. Conductivity to the wellbore may be compromised as branching becomes dominant.

With continued treatment injection, this highly branched dissolution pattern results in uniform, essentially total dissolution in the contact area. This can be highly conductive, of course, but only with impracticably large treatment volumes could penetration to a reasonable distance beyond the wellbore be expected.

Both in the laboratory and theoretically, the optimum combination of treatment injection rate and stimulation fluid reactivity (fluid type) can be determined for a given set of conditions. These conditions include formation type and characteristics. In the field, it is very difficult (if not impossible) to make this determination. This is because leak-off of reactive fluid from the dissolution path, or wormhole, to the surrounding formation really cannot be accurately determined. The difficulty in achieving accurate leak-off measurement exists despite the use of models and laboratory leak-off data.

What is important to know here is that higher, rather than lower, treatment injection rates are preferable. Some degree of retardation is desired, rather than none or too much. Acid can be retarded with viscosity (e.g., gelling or emulsification). Slower-reacting acids such as acetic or weaker stimulation fluids such as chelating agents (EDTA, etc.) have application, as well. Cost can be a limitation.

REFERENCES

1. R. S. Schechter, *Oil Well Stimulation* (Englewood Cliffs, NJ: Prentice Hall, 1992).

2. K. Lund, H. S. Fogler, and C. C. McCune, "Acidization I: The Dissolution of Dolomite in Hydrochloric Acid," *Chemical Engineering Science* 28 (1973): 691–700.

3. K. Lund *et al.*, "Acidization II: The Dissolution of Calcite in Hydrochloric Acid," *Chemical Engineering Science* 30 (1975): 825–35.

4. H. S. Fogler *et al.*, "Dissolution of Selected Minerals in Mud Acid" (paper 52C, presented at the American Institute of Chemical Engineers 74th National Meeting, New Orleans, LA, March 1973).

5. H. S. Fogler, K. Lund, and C. C. McCune, "Acidization. Part 3. The Kinetics of the Dissolution of Sodium and Potassium Feldspar in HF/HCl Acid Mixtures," *Chemical Engineering Science* 30:11 (1975): 1325–1332.

6. H. S. Fogler, K. Lund, and C. C. McCune, "Predicting the Flow and Reaction of HCl/HF Mixtures in Porous Sandstone Cores," *Society of Petroleum Engineers Journal* (October 1976): 248–60.

7. W. E. Kline and H. S. Fogler, "Dissolution Kinetics: The Nature of the Particle Attack of Layered Silicates in HF," *Chemical Engineering Science* 36 (1981): 871–884.

8. D. E. Nierode and B. B. Williams, "Characteristics of Acid Reaction in Limestone Formations," *Society of Petroleum Engineers Journal* (Dec. 1971): 406.

9. B. B. Williams, J. L. Gidley, and R. S. Schechter, *Acidizing Fundamentals*, monograph series (Dallas: Society of Petroleum Engineers, 1979).

10. C. N. Fredd and H. S. Fogler, "Alternative Stimulation Fluids and Their Impact on Carbonate Acidizing" (paper SPE 31074, presented at the Society of Petroleum Engineers International Symposium on Formation Damage Control, Lafayette, LA, Feb. 14–15, 1996).

11. C. W. Crowe, R. C. Martin, and A. M. Michaelis, "Evaluation of Acid Gelling Agents for Use in Well Stimulation," *Society of Petroleum Engineers Journal* (August 1981): 415.

12. L. R. Norman, "Properties and Early Field Results of a Liquid Gelling Agent for Acid" (paper SPE 7834, presented at the Society of Petroleum Engineers Production Technology Symposium, Hobbs, NM, Oct. 30–31, 1978).

13. A. W. Coulter *et al.*, "Alternate Stages of Pad Fluid and Acid Provide Leakoff Control for Fracture Acidizing" (paper SPE 6124, presented at the Society of Petroleum Engineers Annual Technical Conference and Exhibition, New Orleans, LA, Oct. 3–6, 1976).

14. A. Settari, "Modeling of Acid-Fracturing Treatments," *SPE Production & Facilities* 8:1 (February 1993): 30–38.

15. A. N. Barron, A. R. Hendrickson, and D. R. Wieland, "The Effect of Flow on Acid Reactivity in a Carbonate Fracture," *Journal of Petroleum Technology* (April 1962): 409–15.

16. B. B. Williams *et al.*, "Characterization of Liquid-Solid Reactions: Hydrochloric Acid-Calcium Carbonate Reaction," *Industrial Engineering & Chemistry Fundamentals* 9:4 (1970): 589.

17. G. Daccord, E. Touboul, and O. Lietard, "Carbonate Acidizing: Toward a Quantitative Model of the Wormholing Phenomenon," *Society of Petroleum Engineers Production Engineering* (Feb. 1989): 63–68.

18. Y. Wang, "Existence of an Optimum Rate in Carbonate Acidizing and the Effect of Rock Heterogeneity on Wormhole Patterns" (Ph.D. diss., The University of Texas at Austin, 1993).

19. Y. Wang, A. D. Hill, and R. S. Schechter, "The Optimum Injection Rate for Matrix Acidizing of Carbonate Formations" (paper SPE 26578, presented at the Society of Petroleum Engineers Annual Technical Conference and Exhibition, Houston, TX, Oct. 3–6, 1993).

20. C. N. Fredd, "The Influence of Transport and Reaction on Wormhole Formation in Carbonate Porous Media: A Study of Alternative Stimulation Fluids" (Ph.D. thesis, University of Michigan, 1998).

21. C. N. Fredd, "Alternative Stimulation Fluids and Their Impact on Carbonate Acidizing" (paper SPE 31074, presented at the Society of Petroleum Engineers International Symposium on Formation Damage Control, Lafayette, LA, Feb. 14–15, 1996).

22. C. N. Fredd and H. S. Fogler, "Alternative Stimulation Fluids and Their Impact on Carbonate Acidizing," *Society of Petroleum Engineers Journal* 13:1 (March 1998): 34.

23. C. N. Fredd, "Influence of Transport and Reaction on Wormhole Formation in Porous Media," *American Institute of Chemical Engineers Journal* (Sept. 1998): 1933–49.

24. C. N. Fredd and H. S. Fogler, "The Influence of Chelating Agents on the Kinetics of Calcite Dissolution," *Journal of Colloid and Interface Science* 204:1 (Aug. 1998): 187–97.

25. C. N. Fredd and H. S. Fogler, "The Kinetics of Calcite Dissolution in Acetic Acid Solutions," *Chemical Engineering Science* 53:22 (Oct. 1998): 3863–94.

26. C. N. Fredd, R. Tija, and H. S. Fogler, "The Existence of an Optimum Damkohler Number for Matrix Stimulation of Carbonate Formations" (paper SPE 38167, presented at the Society of Petroleum Engineers European Formation Damage Control Conference, The Hague, The Netherlands, June 2–3, 1997).

27. M. L. Hoefner and H. S. Fogler, "Pore Evaluation and Channel Formation During Flow and Reaction in Porous Media," *American Institute of Chemical Engineers Journal* 34:1 (Jan. 1988): 45-54.

28. M. L. Hoefner *et al.*, "Role of Acid Diffusion in Matrix Acidizing of Carbonates," *Journal of Petroleum Technology* (Feb. 1987): 203-8.

29. K. M. Hung, "Modeling of Wormhole Behavior in Carbonate Acidizing" (Ph.D. diss., the University of Texas at Austin, 1987).

30. K. M. Hung, A. D. Hill, and K. Sepehrnoori, "A Mechanistic Model of Wormhole Growth in Carbonate Matrix Acidizing and Acid Fracturing," *Journal of Petroleum Technology* 41:1 (Jan. 1989): 59–66.

31. R. D. Gdanski, "A Fundamentally New Model of Acid Wormholing in Carbonates" (paper SPE 54723, presented at the Society of Petroleum Engineers European Formation Damage Conference, The Hague, The Netherlands, May 31–June 1, 1999).

32. C. N. Fred and M. J. Miller, "Validation of Carbonate Matrix Stimulation Models" (paper SPE 58713, presented at the Society of Petroleum Engineers International Symposium on Formation Damage Control, Lafayette, LA, Feb. 23–24, 2000).

33. M. J. Economides and K. G. Nolte, editors, *Reservoir Stimulation* 3d ed. (Schlumberger Educational Services, 2000).

11 Carbonate Matrix Acidizing Systems And Procedures

Most carbonate acidizing treatments today are conducted above fracturing pressure. However, some damaged wells must be treated below fracturing pressure in order to be stimulated effectively.

As previously discussed, the most common purpose of matrix acidizing is to restore near-wellbore permeability in, or through, a damaged formation zone. Because damage is bypassed, and not directly removed, formation damage in a carbonate acidizing candidate need not be acid-removable, as it does need to be in a sandstone acidizing candidate.

This section discusses conventional matrix acidizing procedures, and alternative and retarded acid methods, additives, and acid placement.

Appendix B also includes some examples of successful matrix acidizing procedures.

CONVENTIONAL MATRIX ACIDIZING

Matrix acidizing treatment procedures are straightforward, as indicated below:

1. Pickling stage
2. Preflush
2. Acid stage (10–300 gal/ft; normally, 25–150 gal/ft)
3. Overflush stage

Pickling stage

Just as in sandstone acidizing, the injection string (production tubing, drill pipe, and coiled tubing) should be "pickled" prior to pumping the acid treatment, if at all possible. Inhibited 5% HCl or special pickling solutions provided by the service companies may be used. Basic pickling acid is 5% HCl containing an iron-control agent and corrosion inhibitor. Pickling acid may also contain aromatic solvent and surfactant. These may be useful if organic deposits and other debris are expected.

Preflush stage

The purpose of the preflush is to remove organic or inorganic scale from the wellbore tubulars prior to injection of the acid stage. Aromatic solvent, such as xylene, can be used to remove hydrocarbon deposits. To remove rust and other inorganic scale, circulating 3-5% HCl downhole is adequate.

Weaker formic or acetic acid mixtures are acceptable alternatives, especially at high temperatures. The preflush may also serve to displace oil from the near-wellbore area to prevent emulsion or sludge formation. Use of xylene, or fresh water containing surfactants, is adequate.

Acid stage

The purpose of the acid stage is to remove or bypass formation damage. The acid stage is usually 15–28% HCl. Treatment volumes usually range from 10–300 gal/ft. Most treatments require 25–150 gal/ft, depending on the anticipated depth of damage and formation porosity. With a formation porosity of 10%, a 60-gal volume of acid per 1 ft of zone is required to fill the porosity to a distance of 5 ft from the wellbore.

However, penetration of acid in a carbonate formation is not uniform. In most formations, the pores are different sizes and shapes. The porosity may be present in the form of vugs, natural fractures or fissures, or tortuous capillary-like pores. Such heterogeneities in the porous structure cause channeling or wormholing of acid through the formation. The effect of wormholing is the attainment of much deeper than expected acid penetration into isolated portions of the matrix, which may be sufficient to overcome skin damage.

High-strength HCl (28%) has some use in deeper damage removal, but for most cases, 15% HCl is adequate. For fracture acidizing applications, 28% HCl makes more sense. If high-strength acid is needed in a matrix stimulation treatment, 20% HCl may be used. Higher acid strengths raise the risks of emulsions, sludges, and even formation of insoluble reaction products under certain conditions. Therefore, highest strength HCl mixtures need not be used unless absolutely required. Their use should be limited to high-permeability formations, such as those encountered in the Middle East.

Organic acid blends and mixtures of HCl and organic acids are useful in higher temperature applications (>300 °F).

Overflush Stage

The purpose of the overflush stage is to displace acid to the perforations. Fresh water is the most common overflush fluid. Filtered crude oil and diesel may also be used in oil wells, but are not preferred because of possible incompatibilities with acid. Nitrogen gas is an effective overflush, especially in gas wells.

RETARDED ACID SYSTEMS

Retarded acid systems can increase acid penetration depth by slowing or blocking acid reaction. They also can reduce acid leak-off rate to the matrix surrounding wormhole channels during their creation, providing deeper penetration and extension of flow channels formed.

The formation of smaller, more branched wormholes, which is promoted by retarding acid, can be beneficial, at least to a certain degree. As mentioned in the previous chapter, excessive branching is not desirable, as this will reduce live acid penetration and flow path conductivity. Retarding acid to the appropriate extent will result in branching, but not at the expense of the formation of dominant, deeply penetrating wormholes.

Acid can be retarded in three ways: (1) retarding the acid with surfactants; (2) adding organic acids and/or reaction products to acid (chemical retardation); or (3) physical retardation.

SURFACTANT-RETARDED ACID

Acid retardation can be accomplished by adding unique, usually oil-wetting, surfactants to acid. These surfactants coat pore surfaces, thereby temporarily preventing or slowing the rate of acid attack on the pore walls. These systems are simple, therefore desirable in that respect. Surfactant-retarded acid is useful in high-temperature applications as well. Both HCl and organic acids can be retarded with surfactant.

CHEMICAL RETARDATION

Adding organic acids or acid reaction products ($CaCl_2$, CO_2) to HCl will also chemically retard reaction. Calcium chloride is useful when acidizing formations containing anhydrite, because $CaCl_2$ decreases the solubility of anhydrite ($CaSO_4$) in HCl. As a result, the quantity of anhydrite reprecipitated when acid spends will be less. This is the only cost-effective application of $CaCl_2$ as a retarder. Adding CO_2 retards HCl by cooling and by changing reaction equilibria and kinetics.

PHYSICAL RETARDATION

Physically retarding acid reaction is accomplished by thickening (viscosifying) the acid used, or including oil-wetting surfactants that block the reaction at the rock surface. Viscous acids include gelled, emulsified, or foamed acid.[1–3] Combinations also can be used. Surfactant-retarded acid can be gelled or foamed, as well. The intent of viscosifying acid is to slow the rate of acid diffusion to the rock surfaces and the fluid-loss rate from wormhole to unreacted matrix. Both work to increase live acid penetration distance.

As in fracture acidizing, the retarded acid system that likely provides the deepest live acid penetration is an oil-external emulsified acid. The high friction pressures experienced with such systems are especially limiting in matrix stimulation, however.

Foamed acids can be effective in improving contact of longer treatment intervals. As in fracture acidizing, most foams are 60–75 quality. The lightness of foam makes it an effective stimulation fluid for damaged gas wells. Like emulsions, pumping foam at high rates is not always possible either.

Gelled acid is used less often than emulsions or foams in matrix acidizing applications. In matrix acid applications, acid is slightly gelled—not to the extent required in fracture acidizing applications. Generally, slightly gelled acid does not provide the degree of retardation possible with foamed acid system or an oil-external acid emulsion.

In cases where increased retardation is needed, slightly gelled acid is less effective than the foamed acid or emulsified acid alternatives. However, slightly gelled acid (1–2% gelling agent) is generally useful, especially in deeper wells where friction pressure with foam or emulsion may be too high. High-temperature gelling agent is itself a common friction reducer in well stimulation treatments.

Currently, the commercial systems available that will result in the deepest penetration of acid are emulsified acid and surfactant-retarded acid. Viscous acids may contain a fluid-loss additive such as oil-soluble resin or polymer for reduced leak-off. The particulate diverters are not effective in fracture acidizing applications, but in matrix treatment, they can sometimes make a difference.

Bear in mind, though, that the need for deep penetration is often overrated. Most conventional acidizing treatments with some degree of retardation or fluid-loss control (viscosity), using ball sealers for diversion, can effectively bypass formation damage within several feet of the wellbore. In any case, simpler acid systems are preferable.

Please see appendix B for example retarded acid treatment procedures.

CARBONATE MATRIX ACID USE GUIDELINES

In 1984, McLeod introduced his sandstone acid use guidelines to the industry.[4] In that landmark paper, he also introduced carbonate acid use guidelines. His guidelines for matrix acidizing of carbonates (with minor modification) are listed in Table 11–1.

Perforating fluid:	5% acetic
Damaged perforations:	9% formic 10% acetic 15% HCl [(a)]
Deep wellbore damage:	15% HCl 28% HCl HCl-organic blends [(a)] Formic-acetic [(b)] Emulsified acid [(c)] Foamed acid [(d)]

(a) *Organic acids may also be mixed with HCl, especially for high-temperature applications.*
(b) *For higher temperature (>250 °F) applications*
(c) *For deeper acid penetration*
(d) *For improved coverage and possibly deeper penetration*

Table 11–1. *Carbonate matrix acidizing use guidelines.*

These guidelines are useful for designing basic matrix acidizing treatments, especially for new wells or for wells in fields with no previous stimulation history.

PLACEMENT

Treatment placement or diversion in carbonate matrix acidizing is more difficult than in sandstone acidizing. This is because of the high solubility of carbonate formations in acid and the formation of channels. This eliminates the use of particulate diverters, such as rock salt, benzoic acid, and oil-soluble resin.

Three effective methods of acid placement in carbonates are:

- Ball sealers
- Gelled acid
- Foamed acid

Ball sealers in combination with higher injection rates provide sufficient diversion in carbonate matrix acidizing through tubing. A more effective method, in general, is the use of alternating stages of gelled (viscous) acid and regular, nonviscous acid. This method can also be used in lower rate treatments as well as those pumped through coiled tubing.

Gelled acid diverter stages accomplish two things:

1. Reaction rates are significantly slowed

2. Fluid injectivity and leak-off decrease in the interval treated as flow resistance is increased due to viscosity increase[5]

The combination of a decrease in acid reaction and an increase in flow resistance increases the tendency of the subsequent treating acid stage to be diverted elsewhere. The process of pumping alternative stages of gelled acid and ungelled acid can then be continued, allowing the ungelled treatment acid stages to enter essentially the entire zone of interest.

The gelled acid diverter stage also can use crosslinked gelled acid, which has higher viscosity. Gelled acid diverters break with time and temperature, although a residue can remain. However, properly designed viscous acid stages, based on service company laboratory testing, should not leave a residue, but should clean up adequately following acid treatment.

Foamed acid, or foam diverter stages, can be effective. It is not always reliable, but it is a clean, essentially nondamaging method in higher permeability formations. Viscous foam can provide temporary diversion sufficient to impart full zone coverage.

In order to place acid in high-pressure zones or where additional weight is required, acid can be weighted with dissolved salts such as calcium chloride. This weighting will allow the acid to reach the desired density, provided that total dissolved solids are well below saturation.

REFERENCES

1. C. W. Crowe, "Evaluation of Acid Gelling Agents For Use in Well Stimulation," *Society of Petroleum Engineers Journal* (Aug. 1981): 415.

2. C. W. Crowe and B. D. Miller, "New, Low Viscosity Acid in Oil Emulsions" (paper SPE 5159, presented at the Society of Petroleum Engineers National Meeting and Exhibition, Houston, TX, Oct. 6–9, 1974).

3. W. G. F. Ford, "Foamed Acid—An Effective Stimulation Fluid," *Journal of Petroleum Technology* (July 1981): 7.

4. H. O. McLeod, "Matrix Acidizing," *Journal of Petroleum Technology* (Dec. 1984): 2055–69.

5. G. R. Coulter and A. R. Jennings, Jr., "A Contemporary Approach to Matrix Acidizing" (paper SPE 38594, presented at the Society of Petroleum Engineers Annual Technical Conference and Exhibition, San Antonio, TX, Oct. 5–8, 1997).

12 Carbonate Fracture Acidizing Systems And Procedures

Most carbonate acidizing treatments are conducted to remove or bypass formation damage, either real or perceived. If damage is not present, fracturing with acid or with proppant should be considered. If damage is present and believed to be very severe or very deep, fracture acidizing or propped fracturing are the preferred options.

DECIDING BETWEEN ACID FRACTURING AND PROPPED FRACTURING

Acid fracturing and propped fracturing are alternative treatments for stimulation of tight or severely damaged carbonate formations. The processes are fundamentally similar. A fracture is created in a rock by injecting a viscous fluid at a rate and pressure sufficient to part the formation. Fracture height is principally controlled by the stress contrasts in bounding rock layers. Fracture length depends upon the height containment and the leak-off properties of the fracturing fluid.

With propped fracturing, fracture conductivity is maintained by propping open the created fracture with a solid material, such as sand or bauxite.

With acid fracturing, nonuniform acid etching (or differential etching) of the fracture face creates lasting conductivity. This is true as long as stable points of support along the etched fracture remain. These hold the channel open and connect to the wellbore following fracture closure.

There are no set guidelines for choosing between acid fracturing and propped fracturing. Historically, the choice often has been based on individual or collective logic. This results from experience with previous treatment response in the same field or under conditions that might be considered similar. Production response is the best criterion for deciding between the two stimulation methods. Of course, cost is also a factor.

Unfortunately, despite great strides, our industry is still not really able to accurately model or predict the outcome of an acid fracturing treatment. Acid fracturing treatments lack the higher degree of predictability associated with hydraulic fracturing with nonreactive fluids. However, knowledge of formation conditions can provide guidance for choosing the type and size of the stimulation treatment method.

Factors that may suggest consideration of a propped fracturing treatment are:

- HCl solubility is low (< 65–75%)
- The carbonate formation is homogeneous (pure limestones)
- Acid reactivity is low (low-temperature dolomites; < 150 °F)
- The rock is weak and/or it has a very high closure pressure, resulting in poor retention of acid-etched fractures
- Permeability is very low, requiring very long fracture length

Factors that may suggest consideration of an acid fracturing treatment are:

- The carbonate formation is predominately naturally fractured, which could lead to propped fracture complications

- The formation is heterogeneous, with porosity and permeability streaks that are conducive to a higher degree of differential acid-etching of the fracture walls

- Formation permeability is good, but formation damage exists

- The well will not mechanically accept proppant

In general, acid fracturing is the more conservative treatment design because proppant is not pumped. The risk of failing to complete the treatment is also much lower. There is no risk of premature screen-out, which can leave the fracturing tubing string full of proppant. Also, there is no risk of the consequences of proppant flowback.

Acid fracturing is also quite often less expensive than propped fracturing, especially in deeper wells. Therefore, acid fracturing should probably be considered first and ruled out before choosing propped fracturing. As Larry Harrington, an industry pioneer in hydraulic fracturing, has said, "Why plug an infinitely conductive channel with proppant?"

Another advantage of acid fracturing is that an acid frac can create conductivity to, but not within, an undesirable sandstone or shale interval. Furthermore, if effective etched conductivity can be imparted, flow turbulence in the fracture would be expected to be less in an open acid fracture than in a fracture containing proppant.

A disadvantage of acid fracturing is that controlling the leak-off rate of reactive acid in a fracture is very difficult. Without the benefit of field experience in a particular formation, predicting etched conductivity and fracture length with a high degree of confidence is not possible. This is due to unknown leak-off characteristics.

Propped fracturing has an advantage for many carbonates. The leak-off coefficient, fracture shape, and proppant conductivity can be estimated or measured with a greater degree of confidence than for an acid fracture. This is because the fluid is nonreactive. A pretreatment data measurement frac (mini-frac analysis) is an established technique used to generate appropriate design parameters for propped fracturing.

In deep formations with high closure pressure, proppant may create a more conductive fracture than can be retained after closure of an acid-etched fracture. This is also true in shallow, soft carbonates. A propped fracturing treatment may create a longer effective fracture length because fluid properties, especially leak-off, are not compromised by reactivity.

In naturally fractured carbonates, propped fracturing may not be appropriate because of the difficulty in placing necessary amounts of proppant. The tortuous paths often present, and the complex stress properties, can result in fracture geometry that is so complex that it becomes impossible to maintain proppant injection.

Natural fractures also can divert the created fracture, causing choke points that are too narrow for proppant to pass. Furthermore, fluid leak-off is invariably high in naturally fractured formations. If it is not accounted for in the frac design, excessive fluid leak-off to natural fractures can result in early proppant screen-out.

Acid fracturing must only be used where good differential etching is probable. The rock strength and closure pressure must indicate that good conductivity will remain after fracture closure.

ACID FRACTURING TREATMENT PROCEDURES

As mentioned previously, there are two general acid fracturing methods or treatment procedures:

- Viscous fingering (pad-acid)
- Viscous acid fracturing

VISCOUS FINGERING

The viscous fingering, or pad-acid method, is one in which the formation is first hydraulically fractured with a nonreactive, high-viscosity gel, normally crosslinked gelled water. This is used to create the desired fracture geometry (i.e., length, height, and width). Next, lower viscosity acid (HCl or an HCl-organic acid blend) is pumped into the created fracture.

The acid "fingers" through the higher viscosity pad because of the viscosity contrast and consequent mobility difference. This phenomenon is called viscous fingering. If the viscosity difference is at least about 50 centipoise (cp), sufficient viscous fingering occurs.[1]

The basic viscous fingering treatment design is given in Table 12–1.

The acid most commonly used is 15% HCl. Higher concentrations, such as 20% or 28% HCl, can also used. Higher HCl concentrations have the advantage of being more viscous than 15% HCl, both initially and after spending. This can help reduce leak-off. HCl/organic acid blends and totally organic acid blends also can be used in place of HCl. Organic acids are useful in higher temperature applications.

Stage	Volume (gal/ft)
1. Acid	100–150
2. Gelled water	100–300
3. Acid w/ball sealers	100–500
4. Gelled water	100–300
5. Repeat steps 3–4 as required *	
6. Acid	100–150
7. Overflush	Acid to perforations

Acid additives required: corrosion inhibitor, iron-control agent

Table 12–1. *Basic viscous fingering design treatment*

Typical volume ranges are listed in Table 12–1. Multistage, higher volume treatments are often called massive acid fracture (MAF) treatments.[2] Acid may be viscous or nonviscous. Most often, it is a good idea to thicken the acid to some extent, especially if HCl is used. Common viscous acid systems are acid-oil emulsion, foamed acid, and gelled acid. However, viscosity contrast between the acid and pad must be at least about 50 cp, as mentioned.

The pad fluid is typically gelled water containing 30–60 pptg (pounds per thousand gallons) of guar or modified guar polymer. The pad may also be crosslinked for higher viscosity and gel stability. The addition of fluid-loss agent and/or 100-mesh sand can be beneficial in controlling leak-off and increasing effective fracture length.

The most effective diversion method in fracture acidizing is with ball sealers. No other method is considered effective, except the natural diversion that may occur as a result of injection of the alternating stages of viscous pad and acid.

VISCOUS ACID FRACTURING

Viscous acid fracturing uses viscous acid systems such as gelled, emulsified, and foamed acid, or chemically retarded acids. These systems are used both to create the fractures and differentially etch the fracture faces. Treatments with viscous acid are applicable in heterogeneous carbonates, such as dolomites or impure limestones.

Viscous acid fracturing has become the more common acid fracturing treatment. The basic viscous acid fracturing treatment design includes:

- Preflush
- Viscous acid stage
- Overflush

Preflush. The preflush is used to initiate a fracture and lower the temperature around the fracture. The preflush is typically slightly gelled (slick) water.

Viscous acid. The purpose of the acid stage is to simultaneously propagate the fracture and differentially etch its walls. The acid stage is typically gelled, emulsified, or foamed acid. Combinations of the three are possible.

As in all carbonate acidizing treatments, 15% HCl is most common. Higher strength HCl, organic acids, and HCl-organic acid blends are also used. Most acid fracturing treatments are conducted with gelled acid. Xanthan gum is an excellent gelling agent for up to 15% HCl. The only problem with xanthan gum is that it does not degrade appreciably at temperatures below 200 °F. It degrades very rapidly—too rapidly—when HCl concentration is greater than 15%.

Most gelled acids use a polyacrylamide gelling agent. Polyacrylamides can be used at low and high temperatures. They can also be crosslinked to attain higher viscosity and gel stability. There are relatively new gelling agent systems that trigger viscosity in situ as acid thins or as acid spends (increasing pH). These systems then thin again as acid spends further to a higher pH, enhancing flowback following treatment.

Overflush. The purpose of the overflush is to displace acid from the wellbore and to push the acid volume forward, thereby increasing the penetration distance. When viscous acid is used, a large overflush can be very effective in increasing the etched fracture length. It is a very important step in the treatment design. A high rate is beneficial.

It is possible to pump plain acid in such a treatment. When plain acid is used, acid reaction is very fast. The acid will dissolve large amounts of rock near the wellbore but create a short penetration distance. If a treatment is designed simply to bypass fairly shallow formation damage, plain acid may be sufficient. If plain acid is used, a large overflush is not needed, because it cannot increase penetration distance. If the intent of the treatment is to stimulate the formation, viscous acid must be used.

More complex versions of the viscous acid fracturing method include alternating stages and alternating acids. With the alternating stage technique, acid and gelled water are alternately pumped.[3] The alternating gelled water stages serve the following purposes:

1. Gelled water stages create greater fracture width because of higher viscosity

2. Gelled water stages cool the fracture, thereby increasing depth of acid penetration (acid reaction is exothermic, therefore localized temperatures in the fracture can become high)

3. The alternating pumping technique helps increase penetration distance if the acid is retarded, because the gelled water helps reduce acid fluid loss from the fracture to the matrix

With the alternating acid technique, two acids with opposite characteristics can be pumped alternately. One acid mixture typically contains reaction-retarding additives. The other acid mixture is nonretarded and will react faster, especially near the wellbore. The idea is to enhance differential etching and to increase dissolution of rock near the wellbore.

The basic viscous acid fracturing method is sufficient for most applications. However, alternating stages or alternating acid techniques have been used very successfully. For stimulating a new well, or a well in a field with no previous acidizing history, though, it is best to keep treatment design as simple as possible.

It is really not possible to definitively predict the outcome of one treatment method in a particular field without prior stimulation history. However, one should not be discouraged by this unpredictability. Successful fracture acidizing is, of course, very possible, and exciting when accomplished. The absence of proppant is a major plus in favor of fracture acidizing.

If desired, a small proppant stage may be added at the end of a fracture acidizing treatment to ensure final conductivity feeding the wellbore. This is not a common practice, but it can make sense.

CLOSED FRACTURE ACIDIZING

Another method for improving final conductivity to the wellbore is the closed fracture acidizing technique.[4] First, it is important to note that the acid fracturing methods previously discussed are successful under most conditions. However, there are formation conditions where the required conductive etched fracture is not developed sufficiently. These conditions include the following:

1. The formation is readily soluble in the acid system used, but the fracture face dissolves uniformly, so that differential etching does not take place. When the fracture closes, conductivity is lost.

2. The formation is etched in an uneven manner, as required, but the etched flow channels are crushed on closure. This is either because the formation is too soft, as in a chalk, or because excessive acid leak-off has softened the fracture face.

3. The formation has a relatively low HCl solubility. When solubility is low, acid-insoluble fines may remain on the fracture face, thereby restricting additional acid reaction needed to create permanent conductivity.

In such cases, a good practice is to include a final matrix treatment stage following acid fracturing. This is called closed fracture acidizing (CFA). CFA is simply pumping a final acid stage at matrix conditions to establish an open fracture to the wellbore, maximizing final inflow.

With CFA, a relatively small volume of acid is injected at low rates below fracturing pressure after the etched fracture created has been allowed to close. By doing so, wide grooves or channels are formed near the wellbore and along the fracture walls or faces. The grooves formed tend to remain open, resulting in good flow capacity under severe closure conditions.

The CFA technique is also applicable as a separate treatment procedure in naturally fractured formations and in previously created fractures, including propped fractures. However, the primary application of closed fracture acidizing is as a final matrix acidizing stage following an acid fracturing treatment.

CFA has been used very successfully following viscous fingering treatments and conventional viscous acid fracturing treatments. Regular 15%

HCl and HCl/organic acid blends are most commonly used as the CFA treatment fluid.

Examples of successful fracture acidizing procedures are given in appendix C.

TREATMENT EVALUATION

It is always important to evaluate the effectiveness of a stimulation treatment. Evaluations of matrix acidizing and acid fracturing treatments are usually based on production increases or comparison with other wells. With matrix acidizing, comprehensive pretreatment and posttreatment well testing and interpretation are usually not economically justified. Comprehensive evaluation can be justified most of the time with acid fracturing and propped fracturing.

Unfortunately, evaluation of an acid fracturing treatment is more difficult than a propped fracturing treatment. This is because of the complexity of fluid leak-off in acid fracturing. With propped fracturing, fluid leak-off of the nonreactive fracturing fluid can be more easily predicted and modeled.

The use of pressure transient analyses to determine fracture penetration and conductivity and formation transmissibility before and after an acid fracturing treatment has been demonstrated.[5]

Evaluation of acid fracturing treatments is complex, especially if alternating fluids are used. In such treatments, the fluid-loss rate and rheological properties of the different injection stages change throughout the treatment. It has been shown that pretreatment and posttreatment data can be used to evaluate complex acid fracturing treatments.[5] Measured data such as temperature profiles, surface pressure, and bottomhole treating pressure and temperature can be used to calculate the fluid efficiency and net fracturing pressure. Fracture geometry also can be determined from this information.

Olsen and Karr of Schlumberger describe the use of measured bottomhole pressures during acid fracturing treatments involving alternating stages of gelled water and acid. A pseudo-three-dimensional simulator is then used to history-match stimulation response.[6] By doing so, various effects of the treatment can be quantified, allowing for future field treatment optimization.

The service companies offer evaluation and interpretation methods that can be well worth the added expense, as long as sufficient data exist.

REFERENCES

1. R. D. Gdanski and W. S. Lee, "On the Design of Fracture Acidizing Treatments" (paper SPE 18885, presented at the Society of Petroleum Engineers Production Operations Symposium, Oklahoma City, OK, March 13–14, 1989).

2. S. W. McDonald, "Evaluations of Production Tests in Oil Wells Stimulated by Massive Acid Fracturing Offshore Water," *Journal of Petroleum Technology* (March 1983): 275.

3. A. W. Coulter *et al.*, "Alternate Stages of Pad Fluid and Acid Provide Leakoff Control for Fracture Acidizing" (paper SPE 6124, presented at the Society of Petroleum Engineers Annual Technical Conference and Exhibition, New Orleans, LA, Oct. 3–6, 1976).

4. S. E. Fredrickson, "Stimulating Carbonate Formations Using a Closed Fracture Acidizing Technique" (paper SPE 14654, presented at the East Texas Regional Meeting of the Society of Petroleum Engineers, Tyler, TX, April 21–22, 1986).

5. C. D. Wehunt, "Evaluation of Alternating Phase Fracture Acidizing Treatment Using Measured Bottomhole Pressure" (paper 20137, presented at the Society of Petroleum Engineers Permian Basin Oil and Gas Recovery Conference, Midland, TX, Mar. 8–9, 1990), 427.

6. T. N. Olsen and G. K. Karr, "Treatment Optimization of Acid Fracturing in Carbonate Formations" (paper SPE 15165, presented at the Society of Petroleum Engineers Rocky Mountain Regional Meeting, Billings, MT, May 19–21, 1986).

Carbonate Acidizing In Horizontal Wells 13

Both matrix acidizing and fracture acidizing have application in horizontal wells completed in carbonate formations. Matrix acidizing is probably more common and more useful in horizontal completions in carbonates than it is in vertical wells completed in carbonate formations. This is because of the long intervals encountered in horizontal wells, and the limitations inherent in fracturing horizontal completions, in general.

A common need in horizontal well stimulation is drill-in fluid damage removal. This damage is typically in and near perforations. Deep formation damage is unlikely, unless solids have been lost to natural fractures. With damage very near the wellbore, matrix acidizing is called for. If deep damage is present, or deep plugging in natural fractures exists, large matrix treatment still may have a chance of success. However, treatment above fracturing pressure may be necessary.

Fracturing may be needed simply to initiate production or raise production rate to an economical level in a low-permeability formation. In such a case, fracture acidizing or hydraulic fracturing would be the choices. Fracturing in a horizontal well, especially with proppant, is not a simple matter. Fracture orientation may be complicated. Design simulation requires good rock data, including principal stress direction and stress differences.

With respect to acidizing horizontal intervals in carbonates, whether it be matrix treatment or fracturing acidizing, treatment placement is the key. Acidizing may be conducted successfully through packers. However, the limitation is the number of packer settings available. In long, horizontal sections, use of special packers may be risky. Solid diverters are sometimes useful. However, poor cleanup and diverter plugging are legitimate concerns, especially in horizontal completions.

A method presented by Economides *et al.* was previously mentioned in chapter 8.[1] Acid is injected through coiled tubing and a nonreactive diverting fluid is pumped through the coiled tubing/production tubing annulus to supply back pressure. This method thereby directs treating fluids into the formation at or below the end of the coiled tubing string. It is also applicable in carbonate matrix acidizing.

A similar method has been applied successfully in fracture acidizing treatments using coiled tubing. Acidizing fluids are pumped through the coiled tubing string, and viscous diverter (optionally containing particulates for fluid-loss control) is pumped through the annulus. Several stages are pumped as the coiled tubing string is drawn from the toe to the heel of the horizontal section.

There is no entirely reliable placement method for carbonate acidizing in horizontal wells. However, with reasonable and carefully selected attempts at placing the treatment, successful matrix acidizing and acid fracturing treatments of horizontal wells are possible. This is true even in very long horizontal open-hole wellbores.

Successful matrix treatment of horizontal completions in carbonates is sometimes accomplished with alternating stages of foam and acid. The first stage pumped is acid, followed by foam diverter. Foam diverter is pumped until pressure increases sufficiently to indicate that diversion is taking place, and that a subsequent acid stage may be pumped. This process is repeated until treatment completion.

High-rate acid fracturing treatments with gel diverter stages have also been used successfully by Dees *et al.*[2] An acid frac procedure shown to be successful in the Austin chalk is as follows:

1. Prepare well for injection down 2.875" tubing and hold annulus pressure during treatment

2. Pump fresh water prepad at about 30 bpm; fresh water to contain 0.025% friction reducer

3. Acidize formation with desired volume of acid and gel diverter pills as follows:

   A. Pump acid stage at about 30 bpm
   B. Pump 2 bbl of fresh water spacer at about 6 bpm
   C. Pump gel diverter pill at about 6 bpm
   D. Pump 2 bbl of fresh water spacer at about 6 bpm
   E. Repeat as desired, eliminating steps B, C, and D on the last stage

Acid stages typically include 10,000 gallons of 15% HCl with additives as follows:

- 0.1% corrosion inhibitor
- 0.1% friction reducer
- 0.1% cationic surfactant (for water-wetting)

Gel diverter pills are typically 1000 gallons of fresh water with alkaline treated refined guar gelling agent that does not hydrate until exposed to acid or temperature. Pills are added according to a treatment design outlined in Table 13–1.

4. Flush acid with a sufficient volume of fresh water with premixed friction reducer to displace acid out of the tubing and/or casing. Pump flush water at about 30 bpm. Fresh water should contain 0.025% friction reducer.

High-rate matrix acidizing treatments have been reported by Tambini to be effective in long horizontal intervals in naturally fractured carbonates.[3] In

Pill Sequence Number	Polymer Added (refined guar)
Pill 1	300 pounds
Pill 2	400 pounds
Pill 3	500 pounds
Pill 4	500 pounds
Pill 5	500 pounds
Pill 6	500 pounds
Pill 7 and additional stages	500 pounds

Table 13-1. *Treatment Design with Gel Diverter Pills*

these cases, a maximum rate, alternating acid method was used. The procedure starts with adequate coverage of the horizontal interval with plain 28% HCl, followed by alternating stages of 28% HCl and gelled 28% HCl. Acid stages are bullheaded. The maximum injection rate below fracturing pressure is maintained throughout treatment.

REFERENCES

1. M. J. Economides, K. Ben Naceur, and R. C. Klem, "Matrix Stimulation Method for Horizontal Wells," *Journal of Petroleum Technology* (July 1991): 854–61.

2. J. M. Dees, T. G. Freet, and G. S. Hollabaugh, "Horizontal Well Stimulation Results in the Austin Chalk Formation, Pearsall Field, Texas" (paper 20683, presented at the Society of Petroleum Engineers 65th Annual Technical Conference and Exhibition, New Orleans, LA, Sept. 23–26, 1990).

3. M. Tambini, "An Effective Matrix Stimulation Technique for Horizontal Wells" (paper SPE 24993, presented at the Society of Petroleum Engineers European Petroleum Conference, Cannes, France, Nov. 16–18, 1992), 325.

part four

specialized remedial treatments

Inorganic Scale Removal

14

As discussed in chapter 3, scales are inorganic deposits that are sometimes acid-removable. Scale formation is associated with water production and can precipitate in tubing, in perforations, and, in rare cases, back in the formation. Contrary to popular belief, high water production rates or water cuts are not required to form appreciable scale. Even with low water cuts, scale can build up slowly over time, eventually creating significant flow restriction.

Scale is most commonly formed in tubulars and perforations, where fluid pressure drops are most acute. Scales form under conditions of lowered pressure and/or reduced temperature, allowing scale to crystallize or solidify out of solution. Therefore, scales can deposit during production and injection of fluids. Also, scales can form when incompatible waters are mixed, independent of pressure and temperature conditions. For example, mixing seawater completion or workover fluid with certain formation brines can precipitate scales.

While scales can be recurring, persistent, and quite troublesome in causing severe restriction to production, they are often relatively easy to contact with stimulation fluids because of their presence in or very near the wellbore. Deep scale damage in a formation is highly unlikely and is rarely encountered.

The most common oilfield scales are calcium carbonate or calcite ($CaCO_3$), calcium sulfate ($CaSO_4$), and barium sulfate or barite ($BaSO_4$). Calcite is the most common and the most easily treatable. Calcium sulfate scale is usually present in the gypsum crystalline form, and is therefore often referred to as gyp scale. Other scale deposits include strontium sulfate ($SrSO_4$), silica (SiO_2) and various silicates, and salts such as sodium chloride or halite (NaCl).

Scale deposits also can include iron compounds such as iron carbonates ($FeCO_3$), iron oxides (Fe_2O_3), iron sulfate ($FeSO_4$), and iron sulfide (FeS). The iron compounds are usually formed from corrosion reactions, rather than naturally through the production of water. Compositions of actual scales can be quite complex, as they may form as combinations of two or more compounds.

Calcite scales can be easily dissolved with hydrochloric acid (HCl). In most cases these scales can be removed, at least temporarily, with a conventional acid treatment. The standard scale-dissolving solution is 15% HCl containing corrosion inhibitor and an iron-control agent. At high temperatures, organic acid, such as acetic, has application as well.

HCl, with its high reactivity toward carbonates, has little trouble doing away with calcite. However, if scale is in the perforations, and beyond in the formation, spent HCl (high-concentration calcium chloride brine) can begin the scale precipitation process over again. Calcium ions (Ca^{2+}) in high concentration may come into contact with water containing bicarbonate ion. If this occurs under pressure drop conditions favorable for calcium carbonate precipitation, calcite can reform at an accelerated rate.

Most iron scales can be treated with acid with varying degrees of success. Greater care must be taken to maintain soluble material in solution. A properly selected iron-control agent is imperative. Iron sulfide is partially soluble, or slowly soluble, in HCl, but that is the only viable treatment option. It may be safer to remove iron scales mechanically by drill-out, milling, or high-pressure jetting operations.

Other scales, such as calcium sulfate ($CaSO_4$) and barium sulfate ($BaSO_4$), are only slightly soluble in acid. Both are partially soluble in certain complexing agents such as EDTA (ethylenediaminetetraaectic acid). EDTA exists in five salt forms; two of the five are practical for use in scale removal. Table 14–1 lists the five salt forms.

	pH	Solubility	Use
H_4EDTA "EDTA acid"	2.0	Low	Iron-control agent in acid, not a treating fluid
NaH_3EDTA monosodium EDTA	3.5	Low	Not used in well treatment (impractical)
Na_2H_2EDTA disodium EDTA	4.7	Moderate	Calcite and sulfate scale removal
Na_3HEDTA trisodium EDTA	8.3	Good	Not used, cost-prohibitive
Na_4EDTA tetrasodium EDTA	10.6	High	Sulfate scale removal, also used as an iron-control agent

Table 14-1. *EDTA Salt Forms*

For the purpose of removing scale, disodium EDTA (Na_2H_2EDTA) and tetrasodium EDTA (Na_4EDTA) are the most effective. They are the most cost-effective, as well, relative to the other EDTA salts. Disodium EDTA is not nearly as cost-effective as hydrochloric acid, of course, but it has other advantages. Unlike HCl, disodium EDTA can dissolve $CaSO_4$. Solubility depends on the crystal structure. EDTA treatments for scale removal also can prevent rapid redeposition of scale.[1, 2]

Calcium sulfate scale can also be treated with gyp converters and then acidized. Gyp converters are water-based solutions that react with the calcium sulfate, converting the scale into acid-soluble components. Sometimes such a treatment need not be followed with acid, as the converted scale may be more easily dispersed. If this is the case, the scale can be removed mechanically; it can either be circulated or swabbed out.

Sometimes it is desirable to avoid acid, especially if the water cut is already very high. Converter treatments are not efficient, and they are often ineffective. High confidence in such treatments would not be warranted.

For chemical treatment of other sulfate scales ($BaSO_4$ or $SrSO_4$), tetrasodium EDTA solution is preferable to other alternatives because of its higher pH (~10.6). The dissolution process is increased at higher pH. Also, tetrasodium EDTA has a much higher solubility limit than the more acidic EDTA salts.

Special and proprietary scale dissolvers also exist. Typically they are high-pH solutions containing chelators or complexing agents, such as EDTA or other similar compounds. It is not clear that such special solutions are significantly more effective than tetrasodium EDTA, or tetrasodium EDTA in caustic solution, for example.

Generally, however, mechanical removal of such scales is much more reliable than chemical treatment. In addition, all forms of scale can be inhibited, at least temporarily, with proper selection and application of chemical treatments. This sometimes can be accomplished in conjunction with an acid treatment to prevent future scaling problems.

To design a proper scale removal or inhibitor treatment, a scale analysis must be conducted. A sample of scale must be obtained and chemically analyzed to determine its composition. Scales can be complex mixtures of sulfates, carbonates, and iron compounds. Consequently, proper identification, or at least prediction, is important in selecting the proper treatment option or combination of options. It is easy to misidentify a scale by assuming an incorrect composition in the absence of a sample analysis.

Table 14–2 summarizes general treatments and preferred solutions for various scale types. Becker provides a technical reference on scale.[3]

Scale	Treatment[a]
$CaCO_3$	HCl (5–15%) Na_2EDTA (up to ~9%)
$CaSO_4$	Na_4EDTA "Gyp converter"
$BaSO_4/SrSO_4$	Na_4EDTA High-pH chelating treatment
Iron oxides	HCl
$FeSO_4$	HCl (15%) Na_2EDTA
FeS	HCl (15%)

Table 14-2. *Scale Removal Treatments*

REFERENCES

1. C. M. Shaughnessy and W. E. Kline, "EDTA Removes Formation Damage at Prudhoe Bay," *Journal of Petroleum Technology* (Oct. 1983): 1783–92.

2. T. N. Tyler, R. R. Metzger, and L. R. Twyford, "Analysis and Treatment of Formation Damage at Prudhoe Bay, AK" (paper SPE 12471, presented at the Society of Petroleum Engineers Formation Damage Control Symposium, Bakersfield, CA, Feb. 1984), 11–22.

3. J. R. Becker, *Corrosion and Scale Handbook* (Tulsa: PennWell, 1998).

15 Organic Deposit Removal

In any discussion about acidizing, organic deposition and its removal must be discussed as well. It is a common damage mechanism, and its possible presence in an oil well that is a potential stimulation candidate must always be explored. This short chapter briefly addresses organic deposition problems and solutions.

Production decline due to downhole organic deposition is sometimes mistakenly assumed to be due to an acid-removable damage mechanism. This can happen if a well is not properly or thoroughly diagnosed. Wax or asphaltene deposition will cause production declines similar to those caused by inorganic scale buildup or near-wellbore fines migration and entrainment in pore throats. Entirely different treatments are required to address these different damage mechanisms or contributors. It therefore bears repeating that formation or wellbore damage must be carefully and properly diagnosed before a stimulation treatment is considered and designed.

There are two general or basic forms of organic deposition: paraffin (wax) deposition and asphaltene deposition. As mentioned in chapter 3, deposition of paraffin, or wax, is a function of temperature. If temperature drops slightly below the cloud point of the oil present, paraffin will begin to coagulate and drop out of solution.

The cloud point is the temperature at which wax deposition begins. This appears first as a faint, cloudy dispersion. A decrease in bottomhole temperature can occur whenever injected or circulated fluids contact formation oil. This can occur during drilling, completion, workover, and stimulation operations.

Acid does not remove paraffin. Paraffin can sometimes be removed with heat or with an increase in temperature (for example, with hot oil treatments). Hot oil treatments, however, are not recommended, as the results are usually quite temporary, when successful. Aromatic solvents, such as xylene or other commercially available substitutes, are more appropriate and are more commonly used to dissolve or at least disperse wax solid deposits. These treatments also meet with temporary and mixed success.

Water-based solutions containing surfactants and perhaps small amounts of naturally occurring solvents are available alternatives to aromatic solvents. They are not as reliable, but may be the only choice in areas subject to stringent environmental restrictions. Other unconventional treatment methods are, from time to time, available commercially. These should be tried only with experimentation in mind. A variety of methods, some very creative (both mechanical and chemical) have come and gone over the years.

Asphaltene deposition is the other major type of organic deposition occurring in oil wells. As mentioned in chapter 3, asphaltenes are heavy-end hydrocarbon components that are suspended as very small colloids in crude oil. These small particles, or colloids, are stable in suspension. Asphaltenes will deposit when flowing oil experiences a sharp pressure drop, destabilizing the suspension.

Pressure drop is highest as oil enters the wellbore or perforation tunnels from the formation. Therefore, asphaltene deposition is most severe at the near-wellbore and in perforations. Asphaltene deposition may also occur when oil contacts certain fluids—especially those with extreme pH (high or low).

Like paraffin deposits, asphaltenes cannot be removed with acid treatment. Acid can cause asphaltene deposition, or more specifically, sludge, as discussed in chapter 3 and chapter 6. For all intents and purposes, sludge damage is irreversible.

Asphaltenes have historically been removed with hydrocarbon solvent treatments. As with paraffin, xylene or other commercial aromatic solvents are most commonly used. There are also certain water-based treatments available commercially. Typically such solutions contain surfactants and possibly small amounts of naturally occurring solvents. They are generally unproven and unreliable. However, if hydrocarbon solvent cannot be pumped because of environmental concerns or restrictions, then there may be no choice but to try a water-based solution.

Both paraffin and asphaltene damage may require repeated treatment or sequences of treatment and flowback. This is because dissolution of deposited solids may be inefficient or slow, requiring repeated efforts. Also, a soak time of perhaps 2–24 hours may be required to slowly solubilize deposits effectively before flowing back. Typical soak periods are from 4–8 hours. The higher the temperature, the shorter soak time required. Laboratory testing can provide guidance in determination of necessary soak time.

Prevention or inhibition of organic deposition is far more effective than removal. There are many commercially available inhibitor treatment products. Selection is based on careful laboratory evaluation and testing of produced crude oil samples.

The most important point that must be made here is that produced oil samples must be analyzed prior to designing a well stimulation treatment. It is essential to determine the presence of paraffin and/or asphaltenes and the potential for deposition. If deposition is likely to occur, or is known to occur, solubility of deposits in the various solvent choices must be evaluated for selection. It is imperative to request that the service company conduct such laboratory testing, not only to select the proper treatment, but especially to avoid improper treatment, such as acidizing.

As discussed in chapter 3 and chapter 6, formation damage must be thoroughly assessed, if possible, to determine damage mechanisms present. It is also important to rule out others that may be suspected, so that incorrect stimulation treatment is avoided.

Wax and asphaltene deposition are formation damage types that, if discovered, should not be treated with acid. If acid treatment is necessary to address other damage types, a proper organic deposit removal pretreatment must be employed. If organic deposition is found to be the primary or only damage contributor, acidizing can be avoided, which is always desirable.

Becker provides a reference on organic deposition and inhibition.[1]

REFERENCES

1. J. R. Becker, *Crude Oil Waxes, Emulsions, and Asphaltenes* (Tulsa: PennWell, 1997).

part five

quality control practices

Acid Treatment Quality Control 16

Incorporating quality control measures during all aspects of an acid job can make the difference between success and failure. Quality control monitoring only during the actual pumping of the treatment is not sufficient. In addition to the pumping stages, quality control steps must also be planned and executed during rig-up of equipment, before pumping, and after pumping.

As mentioned in chapter 6, a number of years ago, George King and George Holman of the Amoco Production Company produced a booklet entitled "Acidizing Quality Control at the Wellsite."[1] Included in that booklet is a section on the steps that should be taken in ensuring quality control of an acidizing treatment. This chapter presents a slightly modified version of the quality control steps presented in that booklet, with added commentary.

QUALITY CONTROL DURING RIG-UP OF EQUIPMENT

A. *Inspect all tanks which will be used to hold acid or water. The tanks must be clean. Small amounts of dirt, mud, or other debris can easily ruin an acid job.*

I once attended a treatment in a well in the Cook Inlet, Alaska, in which the preflush tank (actually a cement tank) contained debris from a cement job pumped earlier in another well on the same platform. The treatment was unsuccessful. It was speculated that cement material present in the preflush tank, and presumably injected into the formation, could have contributed to the failure.

Unfortunately, this potential problem was discovered after the acid job, not before. Inspection of tanks after pumping, conducted in the interest of assigning blame, showed that the nearly empty preflush tank contained material from the prior cement job. This was suspected to be the leading culprit in the stimulation treatment failure.

B. *Make sure the service company has the equipment to circulate the acid tank prior to pumping.*

This is important for the protection of the tubulars and to prevent emulsion problems. Acid corrosion inhibitors and other additives can separate to the top of the tank in as little as 2 hours.

There are many examples of unfortunate consequences resulting from poor acid mixing prior to pumping. One glaring example was discovered during a field-wide stimulation program in high-temperature (>500 °F) geothermal wells in the Philippines, conducted nearly 15 years ago. Severe corrosion problems, in the form of damaged or lost acid injection strings, were occurring in a series of acid jobs.

An on-site investigation of subsequent acid treatments was conducted. This led to the discovery that acid mixtures (15% HCl and 12% HCl-3% HF) containing high loadings of corrosion inhibitor were simply not adequately circulated prior to pumping. The incorrect and unfortunate conclusions drawn previously were that the corrosion inhibitor used was perhaps ineffective, or that loadings were not adequate.

C. *The line to the pit or tank should be laid and ready to connect to the wellhead so the acid can be backflowed immediately after the end of the overflush.*

Sometimes this is left undone, especially when there is bit of a race against the clock to complete a treatment. Advanced planning is always important in order to avoid such oversights. Flowing an acidized well as soon as possible, with minimal delay, is an underrated concern in sandstone acidizing.

Unnecessary shut-in time can lead to reprecipitation of acid reaction products near enough to the wellbore to cause radial permeability damage. This is especially true in cases in which acid was not or could not be displaced far away from the wellbore. Generally, intentional shut-in periods in sandstone acidizing treatments are questionable.

Extended shut-in is usually not a major concern in carbonate acidizing—at least not from the standpoint of unwanted reprecipitation reactions. However, it is always best to prepare a well to be returned to production as soon as possible, if there is no reason to shut the well in or to allow for a soak period. Viscous acid systems, especially gels, should be produced back as soon as possible, following the design polymer breaktime, especially at higher temperatures (>200 °F).

Polymers used to gel acid may thicken, forming solid, overpolymerized or crosslinked gel residue "blobs" or "screwdriver handles" in the extreme. These are very difficult to remove. Retreatment with acid may be required, but such treatment would be undesirable and unreliable.

Soak periods are sometimes necessary in scale removal treatments, or in very slowly reacting carbonate formations, such as low-temperature dolomites.

QUALITY CONTROL BEFORE PUMPING

A. *Check service company ticket to be sure all additives for the job are on location.*

It is not unreasonable, and should not be taken as a personal affront to anyone, to check the service company ticket. It is important to make sure that additives included in the treatment design are also included in the delivery of chemical components to be used and mixed on-site. Additives once in a while fail to show up, and sometimes an incorrect additive, or at least an unacceptable substitute, may be delivered instead of what had been requested or specified.

I once attended an HF acid job in the Gulf of Thailand, at which sacks of ammonium bicarbonate were delivered instead of sacks of ammonium chloride, which was to be used to mix the overflush solution. This error set the job back a day and a half because of the logistical difficulties in replacing the wrong additive with the right one.

In addition, a lower quality corrosion inhibitor was delivered in place of the specified, higher end inhibitor. This was apparently done to reduce inventory of a product that was to be discontinued. The lower quality corrosion inhibitor was known to be effective at the temperature to be encountered. The substitute product contained more solids, though, and should not have been acceptable.

However, in the interest of pumping the job in a timely manner, primarily to meet crew and equipment schedules, the lesser inhibitor was used. Under the circumstances at that time, this situation did nothing to improve an already tenuous relationship between operator and service company.

B. *Circulate the acid storage tank(s) just before the acid is injected into the well.*

It is known that certain additives, particularly corrosion inhibitor, can settle to the top of the acid tank, leaving the body of acid inadequately inhibited. It is important to mix acid on delivery, and continuously, if possible, up to injection. If that is not possible, then acid should be mixed immediately prior to injection.

C. *Check the concentration of acids with a test kit.*

In the late 1970s and early 1980s, Watkins and Roberts of the Union Oil Company of California conducted a survey of on-site acid concentrations measured in tanks.[2] The results were alarming. Wide variations existed in acid concentrations, as well as acid concentration gradients within individual tanks.

Injection of even a moderate amount of an excessively high concentration of HF acid can irreparably damage a sandstone formation, particularly one that is sensitive to HF. Most sands are sensitive to HF concentration to some degree.

D. *Make sure the service company personnel know the maximum surface pressure and stay below that pressure.*

It is important to stay below the maximum surface pressure during injection for equipment safety limit considerations. It is also important in sandstones to stay below fracturing pressure. It may be that changes need to be made as the treatment is being pumped. If any changes are in order, they must be well understood by all personnel beforehand and during treatment as they are made.

E. *Check the pressure-time recorder, or any other on-site treatment monitoring or evaluation mechanism for proper operation.*

Evaluation of an acidizing treatment, especially removal of skin damage or effectiveness of diverter stages, is not possible if pressure devices and monitoring equipment are not working properly. The working order of such equipment should be ensured on-site.

F. *Acid-clean (pickle) the tubing.*

Use 5-15% HCl or a special service company pickling acid solution, if provided. The importance of acid pickling tubing cannot be emphasized enough. If tubing is not pickled, debris, iron scale, pipe dope, and other solids can be injected into the formation at the very beginning of an acid treatment. This can irreparably damage the near-wellbore formation, ruining an otherwise well-designed treatment. If possible, always include a pickling pretreatment.

QUALITY CONTROL DURING PUMPING

A. *Control injection rate. Maintain surface annulus pressure at or below 500 psi during treatment.*

B. *Watch the pressure response when acid reaches the formation.*

The surface pressure should slowly decrease if the rate is held constant, indicating skin removal. If the surface pressure rises sharply or rises continuously for several barrels of acid, the acid may not be removing the damage or may be damaging the formation. Acid injection should be stopped and the well flowed back immediately, no matter what objections there may be. Samples of the backflowed acid should be sent to the service company laboratory for analysis.

C. *Note the pressure response when the diverting agent reaches the formation.*

The surface pressure should rise slightly. If there is no diverter response, more diverter, or a different diverter, may be needed in the future. If diverter response is not noted, the injection rate may be increased, if the pressure limit permits, in order to enhance zone coverage. This should only be considered if it does not otherwise compromise design.

D. *Never exceed the breakdown pressure of the formation in a sandstone acidizing treatment, unless absolutely necessary initially to break down the perfs (with a perforation wash tool), or to bypass severe damage to initiate injectivity.*

I once attended an acid job in a sandstone/shale formation on an onshore, central California well, in which the known fracturing pressure was inadvertently exceeded. The treatment was designed to remove solids suspected to be plugging natural fractures. Several diverting stages were employed to maximize zone coverage and contact. However, once fracture pressure was exceeded, the acid found its path of least resistance. The acid apparently continued to follow this path even after injection pressure was relaxed below fracturing level.

Once fracture pressure was exceeded, injection pressure remained virtually unchanged, suggesting an absence of effective diversion and zone coverage. Posttreatment production results indicated as much, as well. Oil rate was only slightly increased.

E. *In the final flush, make sure acid is displaced from the wellbore.*

The full overflush should be pumped as designed. Spent acid, especially in a sandstone, must be sufficiently displaced from the wellbore to reduce the damaging effects of precipitation of acid reaction products. In carbonate stimulation, it is desirable that acid is fully spent before it flows back, in order to avoid possible corrosion problems upon flowback of return fluids.

QUALITY CONTROL AFTER PUMPING/DURING FLOWBACK

A. *Do not shut the well in after acid injection. Flow the well back to the tank or pit as soon as the flow line is connected.*

As mentioned earlier, it is always desirable to flow spent acid back after treatment in carbonates and sandstones, unless there is a benefit to shutting the treatment in. Shutting in the treatment is generally unnecessary.

Flowback design is very important, but it is often treated carelessly. Flowback immediately following acidizing is not normal production, and if care is not taken, severe damage can be caused during this initial period. Three of the most common problems during flowback are:

- Fines migration
- Reprecipitation of acid reaction products
- Back production and processing of additives

Fines are generated during sandstone acidizing and are carried more efficiently by more viscous or more dense spent acid mixtures. To avoid potentially damaging effects, the flow rate should be set low at first. The flow rate then should be gradually stepped up to planned production conditions after spent acid is produced back.

Reprecipitation of reaction products is inevitable. In sandstone acidizing, the only way to minimize the problem once the acid treatment has been pumped is to return the well to production as soon as possible. The process of flowing spent acid back must be initiated immediately. Shutting in an HF acid treatment is the worst thing that can be done with respect to damage caused by reprecipitation of HF acid reaction products.

Most additives included in acid mixtures are produced back after treatment. They can cause upsets to production equipment and facilities. Depending on the circumstances, spent acid must be collected and properly disposed of. Sometimes, disposal practices meeting environmental standards is not practical. Fluid filtration methods may be an option.[3]

B. *Collect at least three one-quart samples of backflowed acid for analysis. Sample the acid backflow at the beginning, middle, and near the end of the flow. If on swab, get a sample from every other swab run.*

If possible, get a laboratory analysis for:

1. Amount, size, and type of solids
2. Strength of returned acid
3. Total iron content
4. Presence of emulsions and/or sludges
5. Formation of any precipitates (besides iron)

This information can indicate the nature of acid spending, whether overtreatment or undertreatment occurred, and whether iron and potential precipitation of reaction products are a concern. It also can indicate if acid mixtures injected might have been incompatible with formation fluids. It is especially important to patiently evaluate return samples before any possibly incorrect conclusions are drawn about the treatment pumped.

If possible, a complete analysis of acid returns can be made by the service company using spectroscopic methods. Concentrations of the following ions should be determined:

- Aluminum (Al)
- Magnesium (Mg^{2+})
- Chloride (Cl^{-})

- Barium (Ba^{2+})
- Potassium (K^+)
- Fluoride ion (F^-)
- Calcium (Ca^{2+})
- Sodium (Na^+)
- Iron (Fe^{2+}, Fe^{3+})
- Silicon (Si)

Such analysis can indicate what the acid stages "spent" themselves on, and whether the spent acid stages produced back in the expected and desired manner. This information can be important in designing future treatments in the same field. For example, perhaps spent HF returns in a sandstone acidizing treatment showed high Ca^{2+} content. This might suggest the need for a larger acid (HCl) preflush. Or, it may suggest the need to back off on the HF stage.

High Si:Al ratios may indicate reprecipitation of aluminum fluoride salts, suggesting the need to modify the designed HCl:HF ratio. Or the modifications could include lowering HF concentration, or using a special HF acid system designed to minimize reprecipitation of such reaction products.

C. *Take the treatment report and the pressure charts to the office for evaluation and placement into the well file.*

Allow the service company to have copies for future reference, in order to facilitate improvement in subsequent treatment design and execution. Both the operator and service company must make an effort to exchange and discuss all relevant treatment information to give each other the best chance for future success.

In any well operation, including acid stimulation, safety must be included as a most important quality control measure. Safety is the subject of the next and final chapter, so that it may be taken away as the final, and hopefully, everlasting point.

REFERENCES

1. G. E. King and G. B. Holman, "Acidizing Quality Control at the Wellsite," booklet (Tulsa: Amoco Production Research Co., 1982).

2. D. R. Watkins and G. Roberts, "On-Site Acidizing Fluid Analysis Shows HCl and HF Contents Often Varied Substantially From Specified Amounts," *Journal of Petroleum Technology* (May 1983): 865–71.

3. P. B. Hebert *et al.*, "Novel Filtration Process Eliminates System Upset Following Acid Stimulation Treatment" (paper SPE 36601, presented at the Society of Petroleum Engineers Annual Technical Conference and Exhibition, Denver, CO, Oct. 6–9, 1996).

Safety 17

Safety always comes first. It is included here as the last chapter so that it is the last subject read, and hopefully the first subject remembered. Needless to say, job safety is of utmost importance. Service companies, equipment and tool providers, and oil companies have all placed great emphasis and priority on job safety in recent years. Still, there are always improvements to be made.

Unfortunately, even when all safety measures are in place, and each individual on site knows his role, other factors can come into play and cause safety hazards. Conditions of overwork and long hours can result from the need to complete a job or to move a rig to another location by a certain time. These situations, among others, can exact a heavy mental and physical toll, compromising safety and risking lives.

Unfortunately, practices that cut corners and avoid on-site "inconveniences" still pervade the industry as well, particularly in remote locations. There are many examples. Someone may climb up an acid tank ladder in open sandals at a hot, remote jungle wellsite location. Or perhaps elsewhere at a small offshore platform, tanks and pumping equipment are jammed into a crowded space, making it difficult to maneuver.

Maybe someone climbs over pressurized surface lines, rather than walking around them, or leans over an ammonium chloride solution tank to dump in an additive and inadvertently inhales ammonia fumes, and so on. All are unsafe practices. All of these situations create the potential for injury, illness, or even loss of life.

A better understanding of chemical products, mixtures, and their properties and associated health factors and risks should be developed. Great strides have been made in producing and providing the proper information, but, quite unfortunately, the learning step is not taken as seriously, generally speaking.

In any case, with respect to safety, no one on-site should take anything for granted. All are responsible for each other.

As indicated by John Ely in his *Stimulation Engineering Handbook*, the prejob safety meeting is the most important part of the treatment.[1] In this meeting, all safety considerations should be discussed in detail. The main purpose of the meeting is to ensure that each individual present at the job site knows his or her responsibility and what to do at all times during the job.

Ely suggests that one lead service company representative and one oil company representative (to whom the service company leader will respond) should be appointed. These two individuals should make the final decisions at the job site. All others should provide suggestions or point out any errors of which they are aware.

Communication during the job is crucial. All individuals present should be familiar with the service company mode of communication. Service company representatives should report their means of communication to the oil company representative during the treatment. They should plan appropriate action to take if that line of communication fails.

Ely provides the following points for a safe treatment. Some commentary is added:

1. Consider all safety rules.

This could also mean "take all safety rules very seriously," even if some seem trivial or obvious. Ensuring safety requires a certain mind-set—one that consciously places all safety rules and considerations first.

2. Designate a gathering area in the event of a disaster.

This should be established and well understood before commencing treatment. If offshore, it may be on another level of the platform or on a treating boat.

3. Establish a maximum treating pressure.

This is based on wellbore and tubular conditions, surface equipment limitations, or, in the case of matrix acidizing, formation fracture pressure limitations. The desire to bump up injection rate to shorten the job time must be absolutely prohibited.

4. Each person should be assigned a specific responsibility.

If someone is there just to observe, that should be established as his or her only responsibility. Sometimes, one may help by recording manual pressure readings or may need to take on added responsibility if job conditions change or if monitoring equipment fails. However, it should be understood that no one should take on any responsibilities or activities that are not explicitly established beforehand or during a job. For example, observers should not take it upon themselves to climb tanks to check fluid levels, or to take samples, without prior consent from the person assigned the lead (see below).

5. Assign leadership (one person in charge from the oil company and one from the service company).

These two should work together to mutually agree on procedures and responsibilities. Any suggestions should be made to both parties. Both should be kept in the loop on all communication regarding job execution.

6. The oil company representative should designate the persons responsible for each task.

Of course, the service company job supervisor should establish the responsibilities of his own crew. These should be made clear to the oil company representative for his understanding. The responsibilities of other parties present on site should be designated, or approved of, by the oil company representative, with communication of such to the service company lead person.

7. *Make sure a good communication network is set up. Have alternate visual commands.*

8. *Have a fire extinguisher placed on the ground for easy access.*

9. *Have two individuals given the responsibility of transporting injured persons to the nearest clinic/hospital.*

10. *Have a designated vehicle set aside for transporting injured personnel.*

With treatments conducted offshore, or in remote locations, this is often not considered. It must be insisted on, regardless of the added expense.

11. *Do a complete head count of all personnel on location.*

It is also a good practice for the lead persons to have an idea where everyone is at all times. One should inform the leads of their whereabouts, if leaving the site temporarily. This can be especially important offshore, where there are multiple decks.

12. *Set up a gathering area in case an accident occurs.*

This may be a different area than that established in step 2. A disaster, such as an explosion or a major equipment failure or leak, may require evacuation to a safe area. In the event of an accident, the gathering area may, therefore, be different—and more immediately accessible.

Fortunately, safety is taken very seriously by both service companies and oil companies, especially the major companies. It has been increasingly so in recent years. However, in order for you, the individual, to be as safe as possible, as well as keep others safe, you must become aware of all on-site safety requirements and policies. This starts with being aware of your surroundings and what is going on at all times on-site while you are there.

Our jobs are important to us, and so is accomplishing successful well stimulation. As important as these things are to us and our respective organizations, nothing means more than our safety and the safety of others. Hopefully, you will never have the occasion to witness a serious injury or fatality on-site. It is a terrible experience for all involved and concerned. With respect to well operations, in general, and to the subject of this book, acidizing, all else diminishes completely in importance in comparison to safety.

REFERENCES

1. J. W. Ely, *Stimulation Treatment Handbook* (Tulsa: PennWell, 1987).

appendices

Example Sandstone Acidizing Procedures

Listed below are examples of successful general sandstone matrix acidizing treatment procedures for a variety of applications and conditions.

Stage	Fluid	Volume
1. Acid pickling	5–15% HCl + additives	as needed
2. Wellbore Cleanout	Xylene	25 gal/ft
3. Acid preflush	15% HCl + additives	60 gal/ft
4. Low strength main acid	13.5% HCl-1.5% HF + additives	100 gal/ft
5. High strength main acid	12% HCl-3% HF + additives	50 gal/ft
6. Overflush	7% NH_4Cl	150 gal/ft
7. Foam diverter	Nitrogen (N_2) foam	
8. Repeat steps 2–5 as necessary (1 stage per 20'perfs)		
Acid additives: corrosion inhibitor, iron-control agent, water wetting surfactant, mutual solvent (5%), nonemulsifier (optional)		

Table A-1. *Common Gulf Coast Moderate-Temperature Oil Well Treatment*

Stage	Fluid	Volume
1. Preflush	Xylene	20 gal/ft
2. Water diverter	Xylene + surfactants	10 gal/ft
3. Preflush	Xylene	20 gal/ft
4. Foam spacer	Foam (service co. specified)	400 gal
5. Diverter	Foam diverter pill (70% quality)	
6. Acetic preflush	10% acetic + 5% NH_4Cl	25 gal/ft
7. HCl preflush	7.5% HCl	25 gal/ft
8. Low-strength main acid	6% HCl-1.5% HF	50 gal/ft
9. High-strength main acid	10% HCl-2% HF	50 gal/ft
10. Overflush	8% NH_4Cl + 3% acetic + 5% mutual solvent + 0.5% fines-fixing agent (FFA)	150 gal/ft
11. Diverter	Foam diverter pill (70% quality)	400 gal
12. Repeat steps 6–10 as necessary.		
Acid additives: corrosion inhibitor, iron-control agent, water wetting surfactant, mutual solvent (5%), nonemulsifier (optional)		

Table A-2. *Gulf Coast Acid/Fines-Fixing Treatment in Gravel-Packed Wells*

Stage	Fluid	Volume
1. Wellbore cleanup	Xylene + 5% EGMBE	25 gal/ft
2. Acid preflush	10% HCl	75 gal/ft
3. Main acid	9% HCl-1% (or 1.5%) HF	75 gal/ft
4. Overflush	7% NH_4Cl + 5% EGMBE	150 gal/ft
5. Diverter	Nitrogen (N_2) foam	
6. Repeat stages 2–4 as necessary (1 stage per 20' of perfs).		
Acid additives: corrosion inhibitor, iron-control agent, mutual solvent (5% EGMBE), water-wetting surfactant, nonemulsifier (optional)		

Table A-3. *Basis Perforation Damage Removal (Used in Gulf Coast)*

Stage	Fluid	Volume
1. Wellbore cleanout	Xylene	25 gal/ft
2. Acid preflush	10% HCl	80 gal/ft
3. Main acid	7.5% HCl-1.5% HF, or 9% HCl-1% HF	120 gal/ft
4. Overflush	5% NH_4Cl	150 gal/ft
5. Diverter	Nitrogen (N_2) foam	
6. Repeat stages 2–4 as necessary (1 stage per 20 ft of perfs).		
Acid additives: corrosion inhibitor, iron-control agent, water-wetting surfactant, mutual solvent (5%), nonemulsifier (optional)		

Table A-4. *Common Gulf Coast Gravel-Pack Acidizing Procedure*

Stage	Fluid	Volume
1. Wellbore cleanout	Xylene	25 gal/ft
2. Acid preflush	10% HCl	25 gal/ft
3. Low-strength main acid	7.5% HCl-1.5% HF	40 gal/ft
4. High-strength main acid	12% HCl-3%HF	25 gal/ft
5. Overflush	7% NH_4Cl + 5% EGMBE + 0.1% clay stabilizer + 3% acetic acid	75 gal/ft
6. Sand	3% NH_4Cl + HEC + sand	5–10 gal/ft
7. Flush	3% NH_4Cl	5–10 gal/ft
8. Repeat stages 2–5		
9. Displacement	Completion fluid	
10. Gravel pack procedure		
Acid additives: corrosion inhibitor, iron-control agent, water wetting surfactant, mutual solvent (5%), nonemulsifier (optional)		

Table A-5. *Common Gulf Coast Two-Stage Acid/Gravel-Pack Procedure*

Stage	Fluid	Volume
1. Tubing Pickling	Specified by service company	as needed
2. Preflush	5% HCl	75 gal/ft
3. Main acid	3% HCl-0.5% HF	50 gal/ft
4. Overflush	3% NH_4Cl	75 gal/ft
Acid additives: corrosion inhibitor, iron-control agent, mutual solvent (5%), water wetting surfactant,		

Table A-6. *Treatment for Water-Sensitive Short Zones (Not Gravel Packed; < 50 ft).*

Stage	Fluid	Volume
1. Preflush	10% acetic + 5% NH_4Cl	75 gal/ft
2. Main acid	10% acetic –1% HF	75 gal/ft
3. Overflush	5% NH_4Cl (+ acetic, optional) [For gas wells: N_2 instead of NH_4Cl]	75 gal/ft
Acid additives: corrosion inhibitor, iron-control agent, water wetting surfactant		

Table A-7. *Conventional Treatment for Sandstones with High-Chlorite Clay*

Stage	Fluid	Volume
1. Tubing Pickling	Inhibited 5% HCl	as needed
2. Preflush	7.5% HCl + corrosion inhibitor + iron-control agent + surfactant +5% EGMBE	100 gal/ft
3. Main acid	7.5% HCl-1.5% HF + same preflush additives	200 gal/ft
4. Overflush	5% NH_4Cl + 5% EGMBE + water wetting surfactant	100 gal/ft
Pump treatment at maximum rate and maximum pressure (matrix conditions); optional diversion with foam (20–30 ft/stage).		

Table A-8. *Treatment for Very Water-Sensitive Gravel-Packed Wells (Used in Indonesia)*

Stage	Fluid	Volume
1. Wellbore cleanup	Xylene	25 gal/ft
2. Acid preflush	15% HCl + N_2	50 gal/ft
3. Main acid (first)	6.5% HCl-1% HF + N_2	75 gal/ft
4. Main acid (second)	Retarded HF system (service company specified)	75 gal/ft
5. Overflush (first)	5% NH_4Cl + N_2	75 gal/ft
6. Overflush (second)	N_2	
Acid additives: corrosion inhibitor, iron-control agent, surfactant, mutual solvent		

Table A-9. *Oil Well Wellbore Cleanup/Damage Removal: Low Reservoir Pressure, Water-Sensitive Sand (Used in Holland)*

Stage	Fluid	Volume
1. Fluid Displacement	N_2	40 gal/ft equivalent
2. Preflush	15% HCl (70-quality foam)	50 gal/ft
3. Main acid	13.5% HCl-1.5% HF (70-quality foam)	75 gal/ft
4. Repeat stages 2–3 with 50-quality foamed acids.		
5. Displacement	N_2	40 gal/ft equivalent
Acid additives: corrosion inhibitor, iron-control agent, water wetting surfactant		

Table A-10. *Foamed Acid for Gas Wells (Used in Northern California and Other Locations)*

Stage	Fluid	Volume
1. Preflush	10% HCl + 10% acetic	300 gal/ft
2. Main acid	12% HCl-3% HF	300 gal/ft
3. Overflush	3% NH_4Cl + 5% EGMBE	100 gal/ft
Pump at maximum matrix injection rate (7–9 bpm, if possible); no diversion. Acid additives: corrosion inhibitor, iron-control agent, 5% mutual solvent (EGMBE)		

Table A-11. *Bullheaded Mud Damage Removal in Naturally Fractured Formations (e.g., Monterey Shale, CA)*

Stage	Fluid	Volume
1. Preflush	15% HCl + 300 scf N_2/bbl	75 gal/ft
2. Main acid	13.5% HCl-1.5% HF + 300–1200 scf N_2/bbl	100 gal/ft
3. Overflush	3% NH_4Cl	100 gal/ft
4. Diverter	Ball sealers: 50% excess of perfs (1 stage/10 ft, or use spotting tool)	
Acid additives: corrosion inhibitor, iron-control agent, water wetting surfactant, nonemulsifier (optional)		

Table A-12. *Mud Damage Removal in Naturally Fractured Clastics or Shales (with Diverter)*

Stage	Fluid	Volume*
1. Non-HCl preflush	3% NH_4Cl	100 gal/ft
3. Main acid	12% HCl-3% HF	200 gal/ft
4. Overflush	3% NH_4Cl	200 gal/ft
Stages pumped at maximum matrix rate; no diverter, even in long zones. Acid first must be spotted with coiled tubing, then pumped through the annulus at a high rate. * Volumes per ft in a horizontal well would be lower. *Acid additives: corrosion inhibitor, iron-control agent, surfactant*		

Table A-13. *High-Rate Damage Removal Technique (AGIP MAPDIR Method): Low-Carbonate Sands (Also for Horizontal Wells with Large Section)*

Stage	Fluid	Volume
1. Asphaltene Removal	Aromatic solvent + surfactants	25 gal/ft
2. CO_2 preconditioner	CO_2	150 gal/ft (liquid CO_2 basis)
3. Acid preflush	15% HCl + CO_2 (50/50)	50 gal/ft
4. Main acid	12% HCl-3% HF + CO_2 (50/50)	150 gal/ft
5. Overflush	3% NH_4Cl	10 gal/ft
Acid additives: corrosion inhibitor, iron-control agent, surfactant, nonemulsifier, antisludge agent		

Table A-14. *CO_2-Enhanced Acidizing for Wells Producing Heavier Crudes (< about 20–25 API; Developed by Amoco)*

Stage	Fluid	Volume
1. Coiled tubing pickling	Inhibited 5% HCl	as needed
2. Cleaning (optional)	Xylene + 900 scf N_2/bbl	25 gal/ft
3. Acid preflush	15% HCl + 900 scf N_2/bbl	50 gal/ft
4. Main acid	12% HCl-3% HF + 900 scf N_2/bbl	100 gal/ft
5. Overflush	N_2	70 gal/ft equivalent
Acid additives: corrosion inhibitor, iron-control agent, nonemulsifier		

Table A-15. *Example Acid Treatment for Deep, Hot (>300 °F) Gas Well Stimulation through Coiled Tubing (Used in Gulf of Thailand; 50-ft Zone)*

Stage	Fluid	Volume
1. Tubing displacement	6% NH_4Cl	as needed
2. Preflush	7.5% HCl (75-quality foam)	25 gal/ft
3. Main acid	7.5% HCl-1.5% HF (75 quality foam)	25 gal/ft
4. Overflush	N_2	acid to perfs
Acid additives: corrosion inhibitor, 20 gpt foaming agent, iron-control agent, water wetting surfactant		

Table A-16. *Deep, Hot (270 °F) Gas Well Stimulation (Used in Onshore Southern U.S. Dirty Sandstone for Water Block/Clay Damage Removal)*

Stage	Fluid	Volume
1. Preflush	5% KCl	25 gal/ft
2. Acid	Formic-acetic acid (10% formic-5% acetic, or 13% acetic-9% formic)	10,000 gal*
3. Overflush	5% KCl	10,000 gal*
* Regardless of zone height *Acid additives: corrosion inhibitor, iron-control agent, water wetting surfactant*		

Table A-17. *Very Deep, Hot (425 °F) Gas Well Stimulation (Used in Mobile Bay, Norphlet Sandstone)*

Example Carbondate Matrix Acidizing Procedures

Listed below are examples of successful, nonroutine carbonate matrix acidizing treatment procedures. Examples listed are from published procedures used by other operators throughout the world.

Stage	Fluid	Volume
1. Pickle tubing	5% HCl + 0.1% corrosion inhibitor (or same acid below)	as needed
2. Acid break down	15% "Iron-control" acid	small volume to break down perfs
3. Acid	15% "Iron-control" acid	25–200 gal/ft (matrix rate)
4. Flush	water	acid to perfs
Acid additives: 15% "iron-control" acid (sometimes called FE Acid) mixture containing 15% HCl, iron-control additive, and corrosion inhibitor		

Table B-1. *Basic Perforation Breakdown Treatment (Used in the United States, Midcontinent Region and Elsewhere)*

Stage	Fluid	Volume
1. Tubing pickling	5% HCl pickling solution	As needed
2. Acid	15% HCl + 1000 scf N_2/bbl	20,000 gal
3. Flush	Fresh water	As needed
Acid additives: iron-control agent and corrosion inhibitor		

Table B-2. *Basic Matrix Acidizing Treatment in Moderate Temperature (150°F) Formation (Austin Chalk)*

Stage	Fluid	Volume
1. Wellbore Cleanout	Xylene	25 gal/ft
2. Acid	7.5% HCl + formic or acetic acid	300 gal/ft
3. Flush	Fresh water	Displace acid to perfs
4. Overflush	N_2	Variable
Acid additives: corrosion inhibitor, iron-control agent, surfactant; acid may also contain 0.5–1% gelling agent (polyacrylamide type)		

Table B-3. *Basic Matrix Acidizing Treatment in Very High Temperature Carbonate (>300 °F)*

Stage	Fluid	Volume
1. Wellbore Cleanout	Proprietary blend: *solvent, mutual solvent, surfactant*	500 gal
2. Acid	15% HCl	150–300 gal/ft
3. Flush	Fresh water	Displace acid to perfs
4. Overflush	N_2	Variable
Acid additives: corrosion inhibitor, iron-control agent, surfactant; acid may also contain 0.5–1% gelling agent (polyacrylamide type)		

Table B-4. *Basic Damage Removal/Bypass in Older Well (Limestone or Dolomite; Well Stopped Flowing)*

Stage	Fluid	Volume
1. Wellbore Cleanout	Xylene	500 gal
2. Acid	28% HCl + additives	150–300 gal/ft
3. Flush	Fresh water	As needed
Acid additives: 1% gelling agent, 2% corrosion inhibitor, iron-control agent		

Table B-5. *New Well Cleanup Treatment for Short Zones (< 50 ft); High-Temperature Dolomite (Smackover Formation; 300–325 °F)*

Stage	Fluid	Volume
1. Cooling	Water	10,000 gal
2. Acid	Retarded 15% HCl*	2000 gal
3. Spacer	Slick (slightly gelled) water	2000 gal
4. Diverter	Ball sealers	—
5. Repeat steps 2–4 six times.		
6. Acid	Retarded 15% HCl* + 300 scf N_2/bbl	2000 gal
7. Spacer	Gelled water + 300 scf N_2/bbl	1000 gal
8. Flush	Water + 300 scf N_2/bbl	Flush to perfs
**Retarded acid: can be chemically retarded (surfactant retarded); slightly gelled; or foamed*		
Acid additives: corrosion inhibitor, iron-control agent (in addition to retarder additives)		

Table B-6. *Example of Multistage Retarded Acid Treatment in a High-Temperature Well (300 °F) with a Long Zone (300 ft)*

Stage	Fluid	Volume
1. Cooling	Water	10,000 gal
2. Acid	Retarded 15% HCl*	200 gal/ft
3. Flush	Slick water	150 gal/ft
4. Acid	Retarded 15% HCl *	200 gal/ft
5. Overflush	Slick water	150 gal/ft

** Retarded acid: can be slightly gelled, emulsified, or foamed*

Acid additives: corrosion inhibitor, iron-control agent (in addition to retarder additives) + 500 scf N_2 in all fluids

Table B-7. *Example of Retarded Acid Treatment for a Short Zone (50 ft) in a High-Temperature Well (250–300 °F or Higher)*

Stage	Fluid	Volume (gal)
1. Unretarded acid	28% HCl	6000
2. Retarded acid	28% HCl (gelled, emulsified, foamed)	3000
3. Unretarded acid	28% HCl	2000
4. Retarded acid	28% HCl (gelled, emulsified, foamed)	2000
5. Unretarded acid	28% HCl	4000

±This treatment can also be bullheaded at frac pressure (up to 40 bpm)
Acid additives: corrosion inhibitor, iron-control agent (in addition to retarder additives)

Table B-8. *Example of Matrix Treatment for 500-ft Section in a Naturally Fractured Limestone (Bullheaded at Maximum Matrix Injection Rate)±*

Stage	Fluid	Volume
1. Prepad	N_2	100,000 scf
2. Retarded acid	Retarded 15% HCl	1375 gal
3. Spacer	Fresh water + 10% methanol	500 gal
4. Regular acid	15% HCl	250 gal
5. Diverter	Diverter fluid + 20 (1.1 sg) ball sealers	To perfs
6. Repeat steps 2–5 two times		
7. Retarded acid	Retarded 15% HCl	1375 gal
8. Spacer	Fresh water + 10% methanol	500 gal
9. Regular acid	15% HCl	250 gal
10. Flush	N_2	Min. required

Acid additives: *acid retarder (chemical retarder—surfactant), corrosion inhibitor, friction reducer)*

Diverter fluid: 2% KCl + 60 pptg gelling agent (guar type) + fluid loss agent

Example of a chemically retarded acid :

- 20% HCl
- 1.2% high-temperature corrosion inhibitor
- 5.0% formic acid (corrosion inhibitor booster)
- Acid-retarding surfactants
- High-temperature iron-control agents
- 2.5% methanol
- 0.2% fluorosurfactant (flowback enhancement)
- 0.3 gpt friction reducer

Table B-9. *Example of a Retarded HCl Treatment of a High-Temperature (>300 °F) Dolomite Gas Well (22-ft Perforated Zone)*

Example Carbonate Fracture Acidizing Procedures

This appendix lists examples of published successful fracture acidizing treatment procedures. The first section presents some lower temperature (< 200 °F) examples (see tables C–1 through C–6). The second section lists high-temperature treatment examples (> 200 °F; see tables C–7 through C–9). Special emphasis is placed on high-temperature treatments, as they require more thought and creativity.

Low-Temperature Treatments (< 200 °F)

Stage	Fluid	Volume (gal)
1. Pretreatment	Water + scale inhibitor	6000
2. Preflush	Crosslinked gelled water	6000
3. Acid	20% HCl + 20 pptg gelling agent	10,000
	Alternate each 2000 gal acid with 2000 gal gelled water + fluid loss additive	
4. Flush	Water	2000
5. Shut down to obtain fracture closure		
6. Acid (optional)	20% HCl (ungelled)	2000
7. Displacement	Water	Acid to perfs
Acid additives: corrosion inhibitor, iron-control agent		

Table C-1. *Basic Treatment for Low-Temperature Dolomite (110–120 °F; Example is in the San Andres Dolomite, Varying Zone Heights)*

Stage	Fluid	Volume (gal)
1. Pad	Crosslinked gelled water	35,000
2. Acid	15% HCl (nonviscous)	10,000
3. Pad	Crosslinked gelled water	15,000
4. Shut down to obtain closure		
5. Matrix acid	15% HCl	5000
Acid additives: corrosion inhibitor, iron-control agent, *Pad fluid: same as would be used in a propped frac treatment (but without the proppant)*		

Table C-2. *Example of Basic Generic Acid Fracturing Treatment for a Tight Limestone (Viscous Fingering/Closed-Fracture Acidizing Method)*

Stage	Fluid	Volume (gal)
1. Pretreatment	Water + friction reducer	5000
2. Acid	28% HCl (ungelled) *Alternate each 5000 gal acid with 5000 gal gelled water + fluid loss additive.*	40,000
3. Shut down to obtain closure		
4. Acid	28% HCl (ungelled) Pump at matrix rate.	5000
5. Displacement	Water	Acid to perfs
Acid additives: corrosion inhibitor, iron-control agent		

Table C-3. *Big-Volume Treatment Procedure Used in Moderate-Temperature (150–200 °F) Mississippi Chalk*

Stage	Pump Rate (bpm)	Volume (gal/ft)
1. 15% gelled HCl	15	15–25
2. 40 pptg gelled water	25	25
3. 15% gelled HCl w/ball sealers	25	50
4. 40 pptg gelled water	25	30
5. 15% gelled HCl w/ball sealers	25	60
6. 10% emulsified HCl	25	30
7. 40 pptg gelled water	25	150
8. Shut down to obtain closure	—	—
9. 10% emulsified HCl	8	150
10. Gelled diesel (optional)	8	50
11. Repeat stages 9 and 10 two times		
12. Crude oil overflush	5	500
Acid additives: corrosion inhibitor, iron control agent Procedure applicable elsewhere.		

Table C-4. *Viscous Fingering/Closed Fracture Acidizing (CFA) Used in Prudhoe Bay (Zone Heights ~ 70–100 ft; ~ 200 °F)*

Stage	Pump Rate (bpm)	Volume (gal/ft)
1. 15% gelled HCl	15	25
2. 15% gelled HCl w/OSR* and ball sealers	5	100
3. 15% gelled HCl	10	75–100
4. 2% KCl flush	10	75–100
5. Shut down to obtain closure		
6. 10% emulsified HCl (retarded)	1	200
7. 15% HCl w/ball sealers	1	75–100
8. Crude oil overflush	1	200–300
Acid additives: corrosion inhibitor, iron control agent **OSR: oil-soluble resin (fluid-loss control)* Procedure applicable elsewhere.		

Table C-5. *Acid Fracturing Procedure [Also Used in Pruhoe Bay; Shorter Zone Heights (~ 50 ft or less); ~ 200 °F]*

Stage	Fluid	Volume (gal)
1. Cleanout (optional)	Xylene	500
2. Flush	Slick water	2000–5000
3. Pad	Gelled water	10,000
4. Acid	Foamed 15–28% HCl (65–75 quality)	5000–10,000
5. Displacement	Water + N_2 + foaming agent	Acid to perfs
Acid additives: corrosion inhibitor, iron control agent		

Table C-6. *Generalized Common Foamed Acid Viscous Fingering Method*

High-Temperature Treatments (> 200 °F)

Stage	Fluid	Volume (gal)
1. Nonviscous acid	28% HCl + 1500 scf/bbl CO_2	2500
2. Retarded acid	28% HCl chemically retarded/gelled + 1500 scf/bbl CO_2	2500
3. Spacer	Slick water, foamed with 1500 scf/bbl CO_2	2000
4. Repeat 1–3 two times		
5. Nonviscous acid	28% HCl + 1500 scf/bbl CO_2	2500
6. Retarded acid	28% HCl, chemically retarded/gelled + 1500 scf/bbl CO_2	2500
7. Flush (to top ofperf)	Slick water foam (1500 scf/bbl CO_2)	2900
28% HCl; gelled to 40 lbs, also containing surfactant retarder Rates ~15 bpm		

Table C-7. *Acid Fracturing Treatment Design for Gulf Coast Dolomites [250–300°F; 12,000–14,000 ft deep; Long Intervals (300–400 ft)]*

Stage	Fluid	Volume (gal)
1. Cooling prepad	Slick water	10,000
2. Pad	Crosslinked gelled water	50,000
3. Acid	Gelled 13% acetic-9% formic	10,000
4. Viscous gel spacer	Crosslinked gelled water	5000
5. Acid	Gelled 13% acetic-9% formic	10,000
6. Overflush	Gelled water (not crosslinked)	500,000
Acid additives: corrosion inhibitor, iron-control agent *Emulsified acid systems are interchangeable with gelled acid.* *Acid emulsions can be either oil-external or acid-external.*		

Table C-8. *Arun Field; High-Temperature Organic Acid Fracturing*

Stage	Fluid	Volume (gal)
1. Cooling preflush	Water + surfactant + scale inhibitor + friction reducer	15,000
2. Viscous preflush	Crosslinked gelled water (60 pptg HPG, crosslinker, 50 pptg 100 mesh sand + 50 pptg fine mesh fluid-loss agent)	30,000
3. Acid	15% HCl + corrosion inhibitor + surfactant + friction reducer + acid gelling agent + 100 pptg 100 mesh sand + 50 pptg fine mesh fluid-loss agent + 500 scf N_2/bbl	20,000
4. Overflush	Water + surfactant + scale inhibitor + scale inhibitor +friction reducer + 30 pptg 100 mesh sand + 30 pptg fine mesh fluid-loss agent	8000
5. Diverter	Crosslinked gelled water	3000
6. Repeat steps 2–5 four times		
7. Viscous preflush	Same as step 2	30,000
8. Acid	Same as step 3	20,000
9. Overflush	Same as step 4	8000
10. Displacement	Water + friction reducer	6600

Table C-9. *Massive Acid Frac (MAF); Example of Treatment in Tommeliten Field, Norway [Halliburton Procedure: 300-ft or Greater Zones, 10,000-ft depth, 265 °F; Applicable in Large Carbonate Zones]*

Surfactants

Surfactants (surface-active agents) are the most common and most versatile of all acid additives. There are only a few classifications or categories of surfactants, but many variations within them. Surfactants are used to reduce interfacial tension, enhance flowback, break emulsions and sludges, and favorably alter wettability. They also are used to disperse and suspend solids in solution and disperse other additives in oil and water. There are excellent available references on surfactants. One is from Allen and Roberts.[1] As they describe:

> *Surfactants or surface-active agents are chemicals that can favorably or unfavorably affect the flow of fluids near the wellbore and are, therefore, significant in considering well completion, workover, and well stimulation.*
>
> *A surface-active agent or surfactant can be described as a molecule that seeks out an interface and has the ability to alter prevailing conditions. Chemically, a surfactant has an affinity for both water and oil. The surfactant molecule has two parts—one part being soluble in oil and another part being soluble in water. The molecule is thus partially soluble in both water and oil. This promotes the surfactant accumulation at the interface between two liquids, between a liquid and a gas, and between a liquid and a solid.*
>
> *Surfactants have the ability to lower the surface tension of a liquid in contact with a gas by adsorbing at the interface between the liquid and gas.*

Surfactants can also reduce interfacial tension between two immiscible liquids by adsorbing at the interfaces between the liquids, and can reduce interfacial tension and change contact angles by adsorbing at interfaces between a liquid and solids.

The principal mechanism of action by surfactants is the mutual solubility, which results from their dipolar nature (water-soluble portion and oil-soluble portion). Surfactants are classified by the charge of the water-soluble or hydrophilic portion, or group. Typically, surfactants are schematically depicted as shown in Figure D–1, where a circle represents the water-soluble group, and a rectangle represents the oil-soluble group.

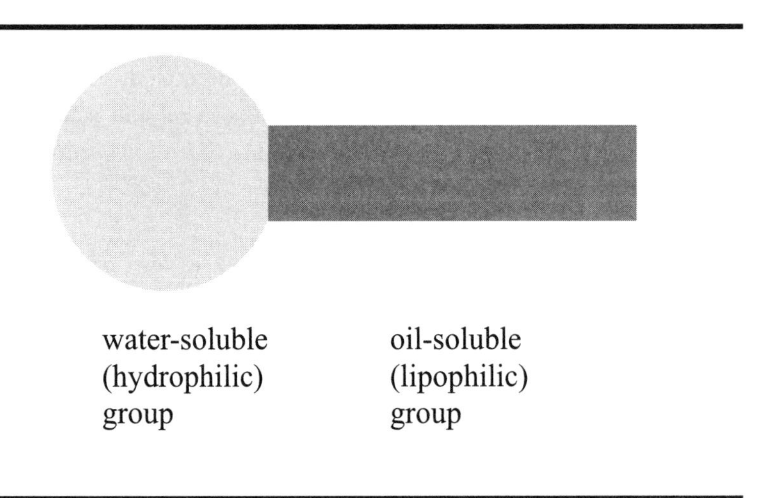

Fig. D-1. *Surfactant*

Surfactants may either be:

- Cationic
- Anionic
- Nonionic
- Amphoteric

Following is a brief summary of these surfactant types and their applications in acidizing, based on discussion originally presented by Allen and Roberts.[1] Ali and Hinkel provide similar information in greater detail.[2]

CATIONIC SURFACTANTS

Cationic surfactants are organic compounds whose water-soluble group is positively charged. This charge is balanced by an anion (X^-), such as chloride (Cl^-). Cationic surfactants are either long-chain amines ($R - NH_3$), or quaternary ammonium compounds, as:

```
      R2
      |
R1 – N – R3
      |
      R4
```

where

R, R1, R2, R3, and R4 are organic chains.

A cationic surfactant can be represented as shown in Figure D–2.

Common applications of cationic surfactants are corrosion inhibitors, nonemulsifiers, and foaming agents.

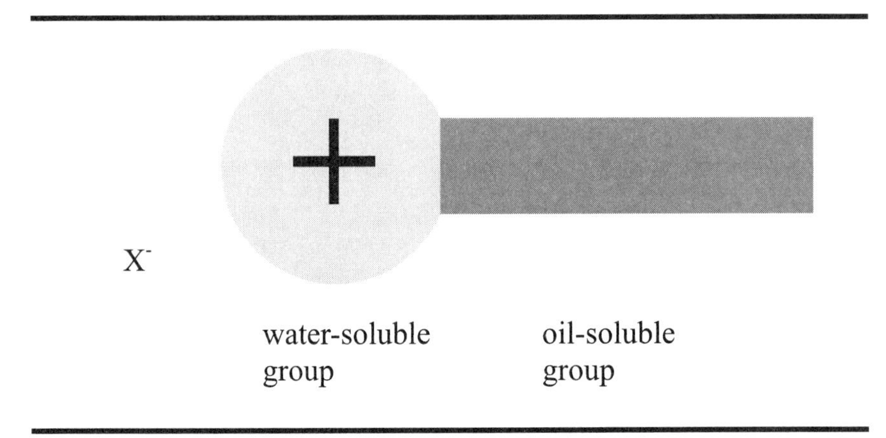

Fig. D-2. *Cationic surfactant*

ANIONIC SURFACTANTS

Anionic surfactants are organic compounds whose water-soluble group is negatively charged. The negative charge is balanced by a metal cation (M^+), such as sodium (Na^+) (see Fig. D–3).

Anionic surfactants are often sulfates ($R - OSO_3$), sulfonates ($R - SO_3$), phosphates ($R - OPO_3$), and phosphonates ($R - PO_3$), where "R" is the oil-soluble organic end. The most common applications of anionic surfactants are cleaning agents and nonemulsifiers.

NONIONIC SURFACTANTS

Nonionic surfactants are surfactants that also contain both a water-soluble group and an oil-soluble group. However, they do not ionize. The water-soluble group is usually a polyethylene oxide or polypropylene polymer. Otherwise, nonionic surfactants include amine oxides and alkanol amine condensates. The oil-soluble end is a long-chain hydrocarbon (alkane). General formulas for the most common nonionic surfactants are:

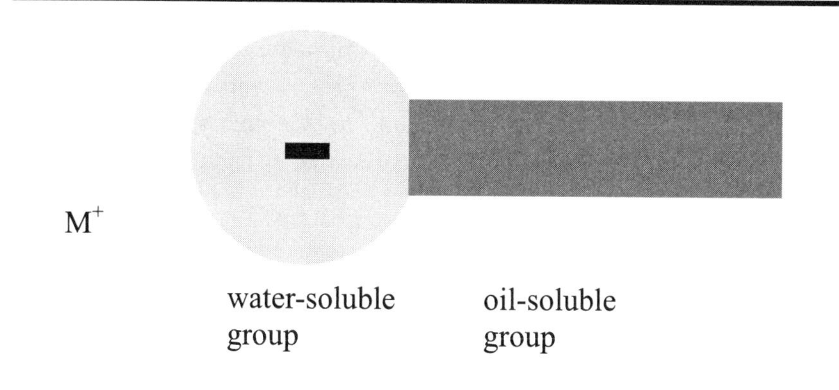

Fig. D-3. *Anionic surfactant*

- Polyethylene oxides: $R - O - (CH_2CH_2O)_x H$
- Polypropylene oxides: $R - O - [CH_2CH(CH_3)O]_x H$

"R" again represents the lipophilic alkane group. The oxide component has an affinity for water and is solubilized as such. Nonionic surfactants are very common, and often most desirable, as there are fewer incompatibility issues with them. Nonionic surfactants are generally compatible with each other and with both cationic and anionic surfactants. Cationic surfactants are often incompatible with anionic surfactants, for example, because of the opposing ionic charges.

AMPHOTERIC SURFACTANTS

Amphoteric surfactants are not common but are occasionally used. Certain foaming agents are amphoteric surfactants, for example. Amphoteric surfactants are organic compounds whose water-soluble group can be positively charged, negatively charged, or uncharged. Examples are amine phosphates, $RNH - (CH_2)_x OPO_3 H$ and amine sulfonates, $RNH - (CH_2)_x SO_3 H$. The R group is the oil-soluble hydrocarbon end.

The ionic charge of the water-soluble end depends on pH of the solution containing the surfactant. Both types (sulfonate and phosphate) have a water-soluble portion made up of two groups with opposite charge. Specifically, charge changes from cationic to nonionic to anionic as pH increases. Amphoteric surfactants are represented in Figure D–4.

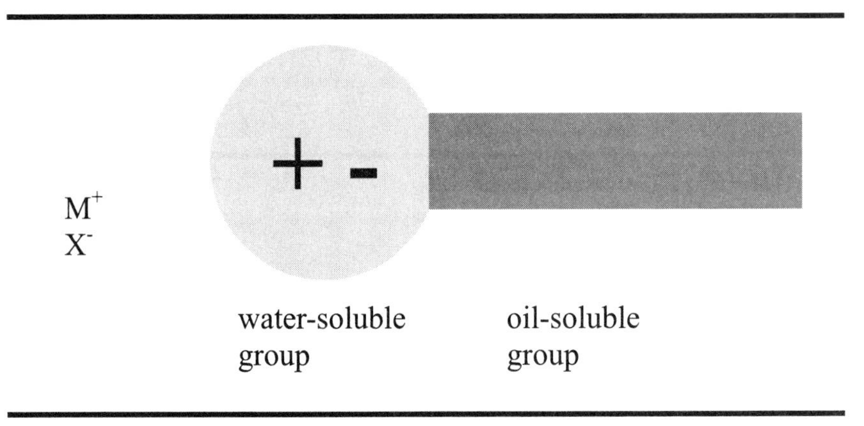

Fig. D-4. *Amphoteric surfactant*

FLUOROCARBON SURFACTANTS

A special class of surfactant is the fluorocarbon type. Fluorocarbon surfactants, or fluorosurfactants, are very effective in lowering fluid surface tension. They are available in cationic, anionic, and nonionic form. Nonionic fluorosurfactants are most common.

Fluorosurfactants are especially useful as flowback enhancement additives and for improving wettability (nonwetting).

REFERENCES

1. T. O. Allen and A. P. Roberts, *Production Operations,* vol. 2 (Tulsa: Oil & Gas Consultants, Inc., 1978).

2. M. J. Economides and K. G. Nolte, editors, *Reservoir Stimulation* 3d ed., (Schlumberger Educational Services, 2000), chapter 15.

Index

A

B

C

D

E

F

G

H

I

K

N

O

P

Q

R

S

T

U

V

W

Z